The Machines of Evolution and the Scope of Meaning

Gary Tomlinson

ZONE BOOKS · NEW YORK

2023

© 2023 Gary Tomlinson

ZONE BOOKS

633 Vanderbilt Street

Brooklyn, NY 11218

Printed in the United States of America.

Distributed by Princeton University Press,
Princeton, New Jersey, and Woodstock, United Kingdom

Library of Congress Cataloging-in-Publication Data
Names: Tomlinson, Gary, author.
Title: The machines of evolution and the scope of meaning /
    Gary Tomlinson.
Description: New York : Zone Books, 2023. | Includes bibliographical
    references and index. | Summary: "Merging recent evolutionary
    thought, theories of information and signs, and new findings
    in animal studies, Gary Tomlinson's The Machines of Evolution
    and the Scope of Meaning offers a groundbreaking account of
    meaning in our world" — Provided by publisher.
Identifiers: LCCN 2022015976 (print) | LCCN 2022015977 (ebook) |
    ISBN 9781942130796 (hardcover) | ISBN 9781942130802 (ebook)
Subjects: LCSH: Meaning (Philosophy) | Knowledge, Theory of |
    Evolution (Biology)
Classification: LCC B105.M4 T7255 2023 (print) | LCC B105.M4 (ebook) |
    DDC 121/.68 — dc23/eng/20220815
LC record available at https://lccn.loc.gov/2022015976
LC ebook record available at https://lccn.loc.gov/2022015977

To Julia, Laura, and Davey

Contents

PART THREE: MEANINGFUL
AND MEANINGLESS COMPLEXITY

In the Realm of Aboutness: Songbirds

Nonsignifying Marvels: Honeybees

Preface

Three conceptual alignments are needed to understand the scope of meaning in the world today. They shape this book.

The first concerns the relation of information and meaning. In casual usage, we assume that information carries meaning. Even in more formal consideration, in the tradition of the mid-twentieth-century invention of information science, meaning is sometimes thought to be equivalent to or coextensive with information. But it is neither; it forms instead a special case, a special kind of information. Information transmission is a crucial aspect of all life-forms, establishing consistent, efficacious relations between them, between parts of them, or between them and their environments. But in this immense network, information bearing meaning is rare. Most transmitted information conveys no meaning. Humans, creatures whose evolved niche-making capacities have left us fated ever to construct meanings, find this counterintuitive, and we need to wrestle in our thought to disconnect information from the imperative of meaningfulness.

I examine the distinction between information and meaning early on in this book, and it forms a basic analytic tool in differences elaborated later. It cuts in two directions. On the one hand, it delimits meaning rather narrowly, locating it in a small part of the whole biosphere and setting borders, or at least border regions, between life-forms that create meaning and others that do not. On the other hand, the same border-setting gesture expands the horizon of meaning beyond *Homo sapiens*, pointing us toward meaning-making processes operating in the negotiations of many nonhuman animals with each other and their surroundings. These processes, like information without meaning, have been hard for us to recognize.

The combined effect of these two directions is to localize meaning as an island in the vast sea of information, but an island that, viewed from the small enclosure humans have erected on it, stretches beyond the easy range of our vision. Gaining the vantage to see the whole island more clearly is part of the work of *The Machines of Evolution and the Scope of Meaning*.

A more distant view glimpses the immeasurable sea of life-forms that live on flows of information without meaning. These are *meaningless* life-forms, in a quite literal sense of that word: life-forms whose informational complexities do not involve or depend on the conveyance of meaning. They are not meaningless in the usual, conversational, evaluative sense of the word that suggests insignificance or unimportance, a sense ringing with ethical overtones and haunted by chain-of-being hierarchies and transcendental constructs such as spirit and soul.

These two senses of "meaningless" introduce another conceptual distinction, that between what we can call processual meaninglessness and evaluative meaninglessness. My focus on processual meaninglessness forms the second alignment of this book — or, better in this instance, *re*alignment, since it amounts to something like a transvaluing of meaninglessness and shifts it from an ethical to an ontological category. This alignment is not an analytic tool, like the distinction of information from meaning, but a set of conceptual implications. In their light, meaning and its opposite are not realms of good and bad, more and less, or important and inconsequential. They don't even necessarily imply gradations of complexity and simplicity but, on the contrary, offer the possibility of spotlighting emergent complexities that have shaped our planet for billions of years, across the entire history of life, *without making meaning*. The transvaluation of meaninglessness positions processes building complexity—from molecular to ecosystemic, cellular to societal levels—such that they cannot seem to measure worth according to humanlike desires, needs, or capacities. A coral reef, jellyfish, paramecium, or butterfly is a piece of meaningless life, but magnificently meaningless in its informational negotiations and dauntingly meaningless in our attempts to understand them.

The repositioning of meaninglessness must also reposition meaning. By virtue of it we come to understand ourselves, along with all other meaning-creating animals, as a swerve, late in evolutionary history, in the mostly meaningless processes governing life. This accident, described in this book as the irruption of processes that allow certain animals to create signs, doesn't set us apart from meaningless life-forms in any fundamental way. It enables striking novelties in the way meaning-makers pursue their homeostatic relations with the external world, but such relations, in a more general view, are universal—all of them transmissions of information, all consequences of the deep, evolutionary machinations by which change in life-forms comes about. Meaning is a small epiphenomenon in the earthly biosphere, a recircuiting of microadjustments in niche management. We can imagine lifeworlds on other planets that, we must presume, have emerged through the operations of the same general machines of evolution active on Earth and the meaningless information they depend on; none of them need have engendered a sign-making machine or its meanings.

There is in this transvalued meaninglessness a grandeur, to borrow a word from the famous close of *The Origin of Species*, though to appreciate it fully Darwin's own humanism, signaled in the same pages, needs to be cast off. In the offloading, the transvaluation points toward strains of thought that have been labeled *posthumanist*, and it is true that the view of meaninglessness here described points to posthumanism in defining retroactively a transhuman commons of meaning-making animals, to which we belong. But its broader lesson goes far beyond the reach of many posthumanist agendas, for it enjoins us to an embrace of the whole biospheric commons, excluding no life-forms, and it does so by nesting the late swerve toward meaning-making processes within the meaningless informational networks that drive all life-forms.

To understand the swerve to meaning and the processes that define it — to understand the meaningless processes in all life-forms that led to meaning-making in some few of them and thus to assay both meaningless and meaningful biotic complexity — requires an approach to evolutionary process that has emerged fully only over the last two decades. This is the *extended evolutionary synthesis*, named to differentiate it from the *modern evolutionary synthesis*, which took shape in the early twentieth century from the entangling of Mendel's genetics and Darwin's natural selection and which dominated evolutionary thought across the rest of the century. *The Machines of Evolution and the Scope of Meaning* is a project in extended-synthesis thinking, but it profits also from theoretical currents from outside that synthesis, especially humanistic ones. This constellation of ideas forms the third conceptual alignment that needs general introduction here, and it calls for a brief detour.

"What I can never understand, in a structure, is that by means of which it is not closed," Jacques Derrida said in 1959, inaugurating with a single sentence the era of poststructuralist thought (Derrida 1978, p. 160). It was *post*structuralist because structuralism had earlier founded itself on a different view of structure or system, rooted in Ferdinand de Saussure's linguistics and semiotics. In this view, meaning was thought to arise from the difference or contrast of units within a structure, the archetypal instance being words in a language. Words mean because they are different from other words in a language; their attachment to things in the world is instituted by that difference, not the source of their meaning. Explaining meaning this way presumes that the structure or system in which it appears is closed, autonomous in its internal differential relations. Followers of Saussure broadened this idea of difference in a closed structure to apply it not merely to language but to cultural systems of all sorts.

They carried his message through many fields, including history, literary analysis, sociology, and, most famously, anthropology.

Derrida's insight was that there can be no structure that is even provisionally closed, since such closure would involve an impossible independence from factors outside a structure determining it. This pushed beyond a mere focus on the openness of many systems, which was not new in 1959, to assert that all structures are perturbed, unsettled, or otherwise conditioned by what is not part of them. They are all open to what lies outside, revealing in their operation the effects of *traces* from beyond — so open, indeed, that words like "outside" and "beyond" lose their customary meanings. What is mooted here — the essential step toward poststructuralism — is not merely a recognition of complex interactions between discrete structures, but *the impossibility of structural discreteness.*

In the human sciences, the stakes of such thinking have been large, its effects broad and irreversible. Identity is seen to be inhabited by its own absences. Being, its boundedness, and its full presence are ceaselessly delayed and relayed, never fixed; self-sameness opens out, dissolving its own enclosure. Meanwhile, interpretation is decentered and reoriented, since any view from within a system cannot grasp traces constitutive of it. Analysis defines structures and their processes through an artificial, freeze-frame view in which the dynamic of traces and the delay of self-identity are invisible. Fixing historical tendencies and trends creates the appearance of narrative closure by smuggling in metaphysical or transcendental terms: origin, goal or telos, and progress. The absolute outside of structures from which these terms are supposed to exercise their control cannot exist. There is no *a priori* place of the sort implicitly granted them.

In destabilizing structures, identities, and presence, Derrida destabilized meaning too, since its structures also are open to the relay and deferral of the trace. Definition of entities must be deferred and understood as a historical process of openness toward the other of whatever is under provisional scrutiny. Poststructuralism's forebears include Nietzsche, who proclaimed, "Only that which has no history can be defined." They also include Charles Sanders Peirce, whose sign theory will play a large role in this book, because its driving impulse is to open semiotic systems radically to what lies outside them, in clear contrast with the closed linguistic systems of Saussurean semiotics.

And, as has become clear in the extended evolutionary synthesis, the forebears of poststructuralism include Darwin too. This has been a slow-to-form, hard-won realization. Open systems were theorized in the sciences long before Derrida, but they were marginalized in the 1950s and '60s, when they ran counter to the systematic aspirations of fields such as cybernetics and early AI, which dreamed of total, top-down control in closed systems. (It took until the 1990s for a thorough, bottom-up reformulation of AI to set in, turning robotics into an enterprise of resolutely open systems.) In late twentieth-century life sciences, the ongoing development of the modern evolutionary synthesis embodied similar totalizing presumptions, even though natural selection had always emphasized the openness of seemingly discrete biological systems such as organisms and populations of them. Evolutionary thought in these decades attempted to establish the discreteness of systems and then arrange their priority; the most famous instance is the "selfish" genome of Richard Dawkins. Theorists debated which systems might be operationally discrete from the "viewpoint" of selective forces (genes? metabolic networks? organisms? populations?) and how selection shaped them. Other evolutionists, more intent on process than structure, asserted other modes of discreteness thought to be generated in natural selection. The key unit here was the *adaptation*, and the construction of historical narratives positing discrete features of organisms, selected to solve discrete existential problems facing them, came to be called *adaptationism* (usually by its critics). In all this, the question of open and closed structures was often raised but not the deeper question of the very possibility of structural closure and discreteness.

Against this background, the extended synthesis, setting off about 1990, represents evolutionary biology's belated owning up to its poststructuralist tendencies. Once again, the stakes are large. In the new view, entities whose stability was once debated but not fundamentally questioned are thrown into doubt: Can we hope to define a species? a gene? Can we delimit biological kinds at all? Terms once confidently held to represent discrete entities are now undone, understood not as innocent, provisional constructs but as distortions of the nature of historical change, rationalist evasions of conundrums in conceiving it. Here the chief culprit is the adaptation. Evolution, according to the extended synthesis, results from structures and systems always set in motion by structures outside them, likewise in motion. Causal arrows are no longer straight but circular. The processes involved in these open structures operate across timescales ranging from the geological to the almost instantaneous, and the far ends of this spectrum, as well as more proximate regions, are looped into one another in networks of efficacy and change.

Despite this new thinking, the extended synthesis has still not fully plumbed its poststructuralist implications. Not that we should hope for an exhaustive mapping of the depths; the *mise en abyme* of current evolutionary thought is activated, as in all productive poststructuralism, by the stimulus of bottomless puzzles in our encounter with the world and time. Still, the depths beckon. Niche-construction theory, for example, one of the chief agendas of the extended synthesis, has already pushed far toward openness, but it needs to take fuller account of momentary, environment-induced changes in gene expression. Intracellular regulatory systems, driving the developmental plasticity of organisms from places beyond DNA itself, hence *epigenetic* places, are assigned various names in the attempt to characterize them. But the names warn us again of our impulse to reify a discreteness that does not exist.

The poststructuralist moment of evolutionary thought, now that it has come of age, swirls with an excitement felt decades ago in other academic neighborhoods. *The Machines of Evolution and the Scope of Meaning* seeks to join this conversation.

Introduction

If in the years since her death the genius of Ursula Le Guin needed quick and certain witness, it would be enough to read her 1974 short story "The Author of the Acacia Seeds. And Other Extracts from the Journal of the Association of Therolinguistics." The first of three imagined scholarly writings it presents interprets a manuscript in Ant, written by a female worker in secreted fluid on acacia seeds and preserved in a deep, isolated tunnel of the colony. Its author, the paper hints, may have been killed by a soldier ant because of her subversive, anti-queen sentiments. The second extract is a call to an Antarctic expedition aiming to expand understanding of literature in Penguin. Its organizer hopes to push beyond conventional readings of penguins' "kinetic sea writings," tackling the most intractable dialect of the language: Emperor, with its midwinter poetry of dark, shared, felt warmth. Finally, in the third extract the president of the association editorializes on language and art, urging members to look past the slow "communicative arts of the tortoise, the oyster, and the sloth" to a new subfield that might decipher the entirely passive, unmoving, and unknown plant arts. Even this new generation of "phytolinguists," he imagines, might not reach the far frontier of art:

> With them, or after them, may there not come that even bolder adventurer — the first geolinguist, who, ignoring the delicate, transient lyrics of the lichen, will read beneath it the still less communicative, still more passive, wholly atemporal, cold, volcanic poetry of the rocks: each one a word spoken,

how long ago, by the Earth itself, in the immense solitude, the immenser community, of space. (Le Guin 2016, p. 625)

From textual interpretation to a call to research action to a neodisciplinary vision, Le Guin frames her panlinguistic fable in deadpan academese hovering between the familiar and the breathtakingly alien. As a fantasy piece, "The Author of the Acacia Seeds" is irresistible.

Today, however, this fable is little resisted not only by sci-fi readers but also in certain scientific and popular scientific quarters. Fantasy seems at times to be confounded with evidence. Our linguistics faculties may not include plant scientists (yet), but we are enjoined to the notions that forests think and that plant cognition is a slow-motion version of its animal counterpart. Phytosemiotics, the study of plant signing, is a recognized subdiscipline in the broader field of biosemiotics. We are told not merely that slime mold growth processes might one day be put to use in computation, but also that the molds themselves solve problems and learn. We are instructed on the numeracy of bacteria. Research on animal communication has moved forward by leaps, with extraordinary new findings in many areas. In the process, however, it has led us to regard as truths about nonhuman experience what were once recognized as metaphors: whale and bird "songs," for example, and animal "speech" and "language." We have grown too comfortable with the transposition of Le Guin's fantasy into scientific assertion.

At what cost? My discomfort doesn't reflect any impulse to minimize the awesome, evolved complexity of even the simplest organisms; this will be clear in what follows. It doesn't stem from a lack of sympathy with animal rights movements and philosophies, which sometimes motivate assertions of humanlike capacities in other organisms. Neither do I wish to oppose efforts to sketch novel, post-human anthropologies aiming to reenvision our place in the biosphere (for two different examples, see Kohn 2013 and Haraway 2016); I see my project instead as allied with those. And I certainly don't want simply to celebrate human exceptionalism — a fact in the world that we cannot help but acknowledge, but whose dire consequences we also see.

The discomfort, instead, is this: I worry that in assimilating to human terms the capacities and behaviors of other organisms we render invisible their own unhuman exceptionalisms. Understanding these depends on our reaching out from our experience toward something else — something related to us but also profoundly alien, like political tracts in Ant, warmth poetry in Emperor, or the lyrics of lichens. This reach, however, must not merely familiarize the alien but also gauge its distance; there is a balance to be struck between assimilation and sheer, never-to-be-assuaged alterity. The trends I shy away from hinder nuanced understanding of difference in order to emphasize undeniably real commonalities, as, for example, when the combinatorial designs shared by birdsong and human language are considered to make birdsong *like* language. Such convergences might more productively engage us as instances of evolutionary histories that have veered toward similar systemic means exploited by radically different organisms in different ways and for different ends. The revelatory significance of each convergence resides in its being at the same time a divergence. It needs to be teased out, its history reconstructed and contextualized in the operation of natural selection and other mechanisms connected to it.

One small branch of this evolutionary history, or perhaps a few small branches, resulted in a set of capacities, shared among a group of animals, that enables them to accommodate the world in a manner different from all other organisms. It allows them to create from their perceptions something we must call, for lack of a less loaded term that will do as well, *meaning*. This book is a study of the conditions under which this creation came into the world. It also involves an attempt to chart the terrain across which the conditions for making meaning pertain or operate, thus mapping the meaningful regions of the biosphere.

This implies that there are other regions of the biosphere, whole large groups of organisms, that are without the capacities necessary to make meaning — regions without meaning at all — *meaningless*. This word names the nonpresence of evolved capacities found in other parts of the biosphere and harbors no value judgment. We have no trouble accepting such nonpresence in countless instances. Most plants can use sunlight, carbon dioxide, and water to synthesize energy-rich carbohydrates; animals cannot. All animals share this lack, this nonpresence, just as they share exceptionalisms involving other, nonphotosynthetic capacities. My first wager in this book is that meaning-making is in a similar same way differentially distributed across living things — that only some animals can make meaning, and that most animals, along with plants and microbes, cannot. Rock poetry makes for alluring fantasy, but our understanding of the differences separating us from other forms of life must expect enlightenment more sporadic than the interpretation of volcanic words.

What it can expect in questions concerning mind, experience, and meaning was usefully circumscribed in another famous writing, philosopher Thomas Nagel's "What Is It Like to Be a Bat?" (Nagel 1974). The punch line of this essay — or the trick played on those for whom the title is a lure, who hope to find out — is that we cannot know. Nagel's article is an entry into debates over the mind-body problem that have been with philosophy since philosophy began, and he makes a principled dismissal of materialists who try to reduce mind to physical properties. He considers the method of reductivism inapplicable to the problem of mind, since it moves to objectivize something that is foundationally subjective. Conscious experience — what it is like to be something (a bat, a human, you, me) — is the first casualty of objective reduction, and its disappearance is fatal to the explanatory power of the method. There is no way to make fully articulate the "is" that links two subjective experiences: what it *is* like for you to be me, for a Martian to be either of us, or for either of us to be a bat.

Solipsism is not Nagel's interest, however. He posits *types* of what-it-is-like-ness, groups with members similar enough to enable them to have more sense of the conscious experience of each other than of members of other groups. The primary type or group in question for him, as for most philosophers, is humans. We can share among ourselves subjective experience more fully than we can do with other creatures, such as bats. But there is a necessary blurriness about Nagel's types, a slippage built into his sharing of experience. To what degree might I share the conscious experience of a chimpanzee, one of a type including humans but not whales? Or of a whale, in a type including humans but not monitor lizards? Or of my dog, member of a group — humans and dogs — in which a mutual domestication has been at work since the Paleolithic period? Surely, in each case, more fully than I can share a flatworm's or houseplant's experience, if any exists.

For me, the group of animals capable of creating meaning forms an extended type in this sense — more extended, probably, than Nagel would allow or find useful. This type came about in particular ways that are explained best by evolutionary theory and the histories it can posit and track. Kinship-in-meaningfulness is what determines the whole type and extends across it. It enables, in variously attenuated degrees, sharing of what-it-is-like-ness and, again in varying degrees, assuaging of alienness. This arises from the play of certain evolved capacities and the processes they sponsor; where they are not present, our sharing of what-it-is-like-ness narrows to the vanishing point. (After his 1974 article, Nagel's own approach to mind took an anti-evolutionary turn — a wrong turn, in my view, based on a simplified understanding of the complexities recent evolutionary theory envisions.)

The overlapping or sharing of conscious experience is precarious, even among humans. Its most robust manifestations are not about the content of experience but about a metaquality, the experience of experience. I can share with you, more than I can share with my dog or a chimpanzee, the experience of experiencing loneliness, but even with you I cannot share the nuance of your loneliness; this is your experience alone. My aim in this book is shaped by this precarity. Except in some very general sense, I don't attempt to understand the content of meanings created by nonhuman animals, but instead to describe the conditions under which they can have content at all, under which meaningfulness arose. Admitting the limit of human subjective consciousness in probing the specificities of other consciousnesses, I try to map a fundamental, evolved space that we share with many other animals, a community of meaning where some metaconscious commonality exists. In addition, and perhaps more intriguingly, I describe how we can look farther afield to understand — objectively, not subjectively, for that would be impossible — the performance of vast behavioral complexity by animals without any meaning at all.

So this short book takes on a tall topic: the sources, nature, and locale of meaning in the earthly biosphere. I have come to this expansive topic gradually, hesitantly. In fact, *The Machines of Evolution and the Scope of Meaning* is the third in a trilogy of books on evolution, with a widening catchment across the three. This trilogy was not planned as such, but with the completion of each book the subject of the next fell into place, calling (as it felt to me) for a sustained effort to make sense of it. From the emergence of musical capacities in the evolution of the hominin clade (*A Million Years of Music: The Emergence of Human Modernity*, 2015) sprang a broader set of issues concerning the evolution of hominin cultures all told (*Culture and the Course of Human Evolution*, 2018). From the putative extension of culture beyond humans in the world today and beyond *Homo sapiens* in the history of hominins, then, came the question of the foundations of culture across many species, foundations located in the creation of meaning.

The Machines of Evolution and the Scope of Meaning is organized into Part I, a preliminary lexicon; Part II, a sustained analysis; Part III, two cases in point; and Part IV, a closing consideration of some questions raised by these. Part I, "Setting Terms," ushers the reader into the fundamental issues that will occupy later parts. After two introductory sections, it is devoted to a description of key concepts, ordered according to a rationale that seems to me logical because it is ontological. *Information*, *mediation*, *sign*, *interpretant*, and *signal* are keywords that distinguish different dynamics in life-forms, setting

24

off the operation of meaning from the broader operation of information transmission. The point of these entries is to stake out positions
on the concepts in question in interaction with other views on them,
but for such general concepts I have made no attempt to be exhaustive; rather, my entries locate and define, in a fairly preliminarily
manner, what I find to be the most useful approaches to the concepts
in question.

Part II forms the analysis. This is the part that engages the
extended evolutionary synthesis at a theoretical level. It describes
and explores four patterns involved in the evolutionary emergence
of meaning that we can think of as *abstract machines*, by which I mean
something so simple as to be almost ineffable: conditions that, if met,
set in motion processes. The first three such machines, natural selection, niche construction, and hypermediated systems regulating the
interaction of organisms with their environments, occupy Sections
4–7. These are foundational for the evolution of all earthly life, and
probably for life anywhere else it may have arisen. At first glance they
seem straightforward, but on closer look they hide mysteries borne
of several features they share: their immense scope and action, their
circular or reciprocal causal pathways, the openness of the structures they bring about, and, perhaps most of all, their immanent,
abstract operation, independent of palpable or material mechanism.
No doubt we can discern within and around these fundamental
designs many others that are essential to life-forms, right down to
very fine granular levels, and many of these rely on quite material
mechanisms. But they are all beholden to and outgrowths of these
abstract machines.

These three machines (natural selection, niche construction,
and mediation) are defining dynamics of all life-forms. From their
machinations over several billion years the conditions arose under
which a fourth abstract machine fell into place: a semiotic machine
that creates meaning. One implication of this statement is crystal
clear: Life on Earth existed for most of its history in forms that did
not generate, process, or understand meaning. The biosphere then

was a rich, changing, abundant locale of meaningless information transmission. To say this cuts against the grain of three current positions among philosophers and evolutionists who argue for a much larger sphere of meaning, and two of these, called *teleosemantics* and *teleodynamics*, form the topic of Sections 9 and 10. I examine these fascinating but unstable positions in the light of the first three evolutionary machines, arguing that their extension of meaning through the whole biosphere, and even beyond it, is unwarranted. The third position that similarly extends meaning, the *biosemiotic* position, is not directly addressed here because an alternative view of semiotics and its extent pervades the book. Indeed, to advance a semiotic view of meaning different from the semantic universalism of most biosemioticians is one of the primary purposes of this project.

The semiotic machine — the fourth abstract machine and the topic of Sections 11–14 — is the linchpin of the book, for the signs it generates are the source of meaning in the world. The semiotic process here is described starting from humanistic theories of it, but these work not simply as heuristic constructions. Instead they characterize outcomes of the workings of the first three evolutionary machines that brought into the world, in a certain group of life-forms, new processes of information transmission. To mark this semiotic machine as a fundamental development in evolution is not customary among biologists, but it calls out to be ranked among the "major transitions" that now form an important part of their thinking — to be counted, that is, as one in a small series of major branchings in the tree of life, in each of which some life-forms abruptly came to manifest unprecedented features. The falling into place of semiosis, in the wake of the formation of certain capacities of certain living things, accords well with biologists' understanding of other such transitions.

The analysis of semiosis here resonates with much recent thinking in its reliance on the ideas of Charles Sanders Peirce. But it differs from most neo-Peirceanism in the particular realm it discerns for sign-making: a realm that is much narrower than that imagined by biosemioticians, even as it is very much wider than the realm of signs proposed by the human-centered semiotics of many other recent thinkers. Peirce himself wavered across the career of his itinerant thought, veering usually toward anthropocentrism but occasionally toward its polar opposite, something closer to panpsychism. There are strong reasons, which I develop, to think that the sphere of signs is delimited within a middle ground between these extremes. Meaning is not exclusively human but instead is the outcome of a process inevitable under certain evolved conditions. These conditions extend far beyond humans, if not very far across the whole biosphere.

I resisted, at the outset of this introduction, posing the kind of question that often begins books on large topics: "Why do we need another book on *X*?" There are indeed many books on meaning from an evolutionary vantage point. Precursors in this area include Ruth Garrett Millikan, who approaches the question as an analytic philosopher of "naturalistic" bent; Daniel Dennett and Kim Sterelny, who emphasize two distinct interpretations of evolutionary theory, which we can respectively denote for now, in terms that will be clarified later, as "adaptationist" and "niche constructive"; and Terrence Deacon, an evolutionary philosopher and practicing scientist, who adds to the picture bracing doses of complexity and emergence theory. (Deacon is also a leading neo-Peircean, and his connection of Peirce's ideas to human evolution in work of the 1990s was a formative stimulus in my own thinking about Peirce.) My indebtedness to and differences with most of these writers will be dwelled on in what follows; it is the differences that justify my own effort.

Put in the most general terms, these concern the need to stake out and provisionally survey a territory of meaning that is neither so broad as to make meaning synonymous with information transmission nor so narrow as to make meaning a marker of human uniqueness. Human uniqueness in the world today is not predicated on the construction of meaning itself, but on more specific outgrowths of it — and even these outgrowths are not uniquely human, only hyperdeveloped by humans.

What are the evolved capacities of animals that found sign- and meaning-making? How do they define the borderline between semiotic and nonsemiotic animals? To chart this border along its full length would expand this book to impossible length, taking me far beyond my expertise into the cases of countless species, but exemplification at least is possible. This comes in Part III and takes the form of studies in the recent science on songbirds and honeybees. I hope to convince the reader that this science, understood in the light of the semiotic machine, shows songbirds to be fully semiotic creatures and honeybees to be nonsemiotic.

I've chosen the case of bees exactly because it seems to pose a hard challenge to my position. The so-called waggle dances honeybees perform have been accepted as an exemplar of eloquent, meaning-laden animal communication ever since they won their interpreter, Karl von Frisch, a Nobel Prize in 1973, and still today they lead researchers to ascribe language, symbolism, and meaningful communication to bees. The science of honeybee sociality, however, has moved far since von Frisch's time, and a careful reading of it says something different about the waggle dances. They are a wonder of evolved social behavior and information transfer, forming part of the larger, wondrous complexity of insect superorganisms; but they involve no signs and they convey no meanings. They are richly efficacious informational mechanisms of beautiful intricacy — and exactly meaningless. Birdsong offers a contrasting case, also brightly illuminated by recent research, that sits on the other side of the semiotic divide. Here we witness a panoply of signs and meaningful behaviors, deployed across thousands of species in as many rich social circumstances. Two implications of the contrast of honeybees and songbirds are already clear in this summary: *neither complex sociality nor intricate communication need emerge from or generate meaning.* We must reserve room in each for both meaningful and meaningless varieties.

The difference between these two cases is sheer, even vertiginous, and it can be followed today down to very fine levels of biological mechanism — neuronal levels, at least, and sometimes even molecular ones. It marks, as I've suggested, opposed positions in relation to the threshold of an evolutionary major transition. This marking returns us to one of the keywords introduced in Part I: mediation. The distinction between information in general and the special case of it that is semiotic arises from differences in the kinds and degrees of mediation required for each. These differences begin to map out the small sphere of meaning constructed on and within the vast realm of information.

I end in Part IV with a series of "Outstanding Questions"—not so much a conclusion as a set of openings. Here broad issues are weighed and adjusted, and some tentative answers suggested, in the light of the fourth evolutionary machine. The topics taken up include several concerned generally with semiosis: the evolutionary history of major transitions, the reach of semiosis, and the nature of the borderline between semiotic and nonsemiotic information. The questions also concern two major outgrowths of semiosis: technology and culture. These latter questions once more underscore the difference brought about by signs in thousands of kinds of animals, even as they isolate this manifold of lifeways to one small corner of the biosphere.

PART ONE

Setting Terms

What key concepts enable us to approach
and delimit the realm of meaning?

ONE

Limits of Transspeciesism

From which direction will we view meaning? The reach of meaning looks short, viewed from the vantage of the whole biosphere; it looks long, if we take its measure from the human enclosure. Such is the position on the scope of meaning I develop in this book. Other options on both sides of this middle ground have their advocates, and the advocacy often tends to extremes: a view from the biosphere that discovers meaning everywhere and a view from our enclosure that sees, beyond it, little or no meaning. The former position we can call *semantic universalism*, the latter *humanist parochialism*.

Researchers in many fields — philosophy, linguistics, psychology, cognitive studies, and, implicitly, the humanities as a whole — investigate meaning as a phenomenon that is mainly or exclusively human. Often this is signaled by a vocabulary that simply sounds human, binding meaning to ideas, concepts, and the like. Whatever capacities we associate with a lion or an egret, conceptual knowledge seems to betoken something foreign to either. As a model for meaning, parochialist positions often feature human language: its words, its propositional structures, its composition of large discrete units (sentences) out of smaller ones (words), and its algorithm-like rules. This linguocentric approach tends to limit meaning to animals with natural language, which is to say humans, and so enforce human exceptionalism. In extreme form, it posits a language of cognition, commonly called *mentalese*, and argues that cognition only occurs as propositions or sentences, or something structured much like them. Then not merely conceptual meaning but cognition itself can come to seem the exclusive province of humans.

Surely, however, cognition pertains to lions and parrots and dogs; surely meaning has a wider catchment than we can see if we limit it to advanced concepts or propositional language. A simple historical thought experiment confirms that the recognition of human *distinctiveness* in the creation of meaning need not lead us to presume human *uniqueness* there.

Homo sapiens is the sole surviving representative of the hominin clade, but think for a moment of other extinct groups in it: our direct ancestors in Africa, our cousins the Neandertals in Eurasia, the Denisovans of southern Siberia, the older *Homo heidelbergensis* (probably the common ancestor of Neandertals and us), still older *Homo erectus*, *Homo ergaster*, and others in a lineage stretching back at least three million years. The members of these different groups, abundant archaeological evidence shows, were complex creatures with varying degrees of technological proficiency, cultures of knowledge passed from generation to generation, and elaborated social interactions. At the same time, most or all of them, inferences from the evidence argue, did not possess human language in its modern form. Would it make sense to propose that none of them created anything in response to their perceptions of the world akin to the meanings that modern humans create? No; it would be absurd to cut off the realm of meaning arbitrarily at the sapient borderline, including modern humans but no ancient ones. This is most obvious in regard to the Neandertals, advanced, resourceful humans with whom sapients occasionally interbred; but the borderline where meaning ends no doubt sits far beyond and before them. So here we are faced, to start, with a history of transspecies dispersion of meaning-creating capacities that humanist parochialism obscures or denies.

Once meaning breaches the species border, we must ask how far it extends. Circumscribing meaning at the sapient border among animals alive today omits all nonhuman great apes, and we have learned too much about their behavior for this to seem a true or likely measure. A more capacious understanding of meaning is required. But if we include apes, what about cetaceans (whales, dolphins, and porpoises), several groups of which show complex, learned, and referential communicative codes? And, from a nonmammalian taxon, what about songbirds, with their combinatorial song structures? Farther afield are cuttlefish, which communicate through arrangements of polarized light (we see it as color, but they don't) flashing across their skin. Might this not convey some kind of meaning? A transspecies and transhistorical perspective opens the prospect of a wide swath of meaning cutting through the biosphere. In the face of it, humanist parochialism is hard to sustain.

The other pole, semantic universalism, carries meaning — and related concepts such as reference, signs, and thought — beyond mammals and birds and even outliers like cuttlefish and octopuses. We begin to lose sight of any special capacities identifiable in a large group of complex animals. Some ecologists and botanists, for example, maintain that thinking takes place and meaning is generated in plants and forests, organisms without central nervous systems (Pollan 2013). Other researchers in the field of *biosemiotics* trace signs, and hence the meaning bound up with them, all the way to genetic transcription mechanisms involving DNA and RNA. Their position was summed up by Kalevi Kull, who wrote in 2009 that "it has become widely accepted within biosemiotics that the semiotic approach is an appropriate tool to describe all living systems, down to the first cells" (Kull 2009, p. 12; see also Hoffmeyer and Stjernfelt 2016).

This way of thinking has entered into research areas more mainstream than biosemiotics. John Maynard Smith, for example, an important evolutionary theorist of the late twentieth century, defined meaning in molecular terms, deeming it to be present whenever genes or proteins function "in a way that favors the survival of the organism" (2000, p. 179). The very relation of genotype to phenotype is considered meaningful and "semantic," so in this view meaning comes to encompass all adaptive outcomes of natural selection. Other biologists have followed suit, asserting that meaning or semantic content inheres in genetic transcriptional machinery — DNA and RNA together with the molecules that regulate their interactions (Haig 2020) — or in the relations of all organisms with their environments (Odling-Smee, Laland, and Feldman 2003).

Semantic universalism has also found prominent advocacy in certain philosophical approaches to life and evolution over the last thirty years. These approaches to meaning are loosely grouped under the name *teleosemantics*, and they are noteworthy for their attempt to discover the evolutionary course from which meaning arose. But I don't think they get their evolutionary theory quite right, and in Section 9 we will see why. We will also take up the most dramatic attempt to widen the range of meaning, Terrence Deacon's *Incomplete Nature: How Mind Emerged from Matter* (2012b), which expands teleosemantics into an embracing thermodynamic theory of emergent, living complexity that Deacon names *teleodynamics*. As their names suggest, both teleosemantics and teleodynamics espouse a directedness in evolutionary process that calls for close examination.

In these universalist positions on meaning, what seems at the pole of humanist parochialism to be a distinctly human capacity has been scattered across all life-forms and all the mechanisms they comprise. The extension seems to me as dubious a position — as unproductive of knowledge and understanding — as is the assigning of meaning to *Homo sapiens* alone. Of course, we are free to define terms as we choose. But the understanding we seek rarely lies exclusively in the view from afar, without granular details — for example, in the view of meaning as a thing spread across the spectrum from human thought to RNA transcription. It almost always requires also gauging and analyzing differences among those details, dividing the broad picture and considering the relations of its parts.

The middle course between humanist parochialism and semantic universalism I advocate in this book forms a limited transspeciesism in which meaning shapes experiences and behaviors across a spectrum of kinds of organisms but is absent from a far larger group. This course implies a borderline — perhaps blurry but nonetheless real and important — between organisms that live out their lives within the sphere of meaning and those that don't. To say this is not to diminish the awesome complexity of organisms without meaning or the complexity of the behaviors they generate, which very often fool human observers, inevitably biased in this regard, into seeing them as full of meaning. Instead it is, on one side, to understand the almost miraculous way in which evolution has shaped such complexities without any role for meaningful experience, and, on the other side, to historicize the emergence of meaning-generating processes in the smaller array of organisms that *do* experience meaning.

We cannot understand this borderline, the different capacities on either side of it, or the location of meaning in the biosphere either by examining uniquely human modes of meaning or by extending meaning by fiat all the way to molecular interactions. Instead we need to consider the evolution and nature of capacities of nonhuman organisms and to see the ways these situate their bearers differently in relation to their environments. These differences enable us to stake out the borderline between meaning and nonmeaning.

Atoms of Aboutness

To historicize the emergence of meaning-generating processes, I have written. Meaning is process for the organisms that create it, a time-bound dynamic of self and surroundings. Meaning is historical, produced by momentary operations of organismal structures in their environments but also tied to the evolutionary histories that have generated these structures and their potentials for processual interaction. Meaning has a history on Earth. It had a beginning in the development of earthly life, and it will have an end.

Histories of meaning need to be connected across hugely different timescales, from the *longue durée* shifts in the most stable evolved features of life-forms to the fleeting changes brought about in individual organisms by life experience. To make these connections we need a small lexicon of basic terms or keywords, defined here so as to orient us preliminarily, then applied and developed as we progress. The terms described in Section 3 encompass large areas of thought and have been varied in their applications. My versions of them will be sensitive to these rich usage traditions without making claims to comprehensiveness. These are tools for us, not entities fixed once and for all.

Key among these keywords are *information*, *sign*, and *interpretant*. The first two admit of many uses, colloquial as well as formal. On the formal side, the distinction I draw between them — such that all signs are informational, but not all information involves signs — is required, I believe, for any natural history of the emergence of meaning and any understanding that would avoid the arbitrary limitation of humanist parochialists or the terminological mission creep of semantic universalists. The distinction involves the third keyword in the group, *interpretant*, coined by philosopher Charles Sanders Peirce in the late nineteenth century, often cited and often misunderstood. Rightly assembled, this triad of terms offers a toolkit to identify and describe the transition in the history of earthly life that brought about meaning.

Signs are of central importance in my account, forming a necessary starting point for any meaning in the world. Signs are always signs of something. They are referential, referring or pointing to something other than themselves and thus bearing what can be called a content. They represent, if we can use this problematic word precisely to mean the re-presenting or presenting over again of one thing in another. They are semantic, in a way not automatically associated with human language, and imbued in their very structure with a quality of *aboutness*. Since a meaningful state of experience or mind is always about something — it points toward or concerns something and is therefore *intentional* in the philosophers' sense of that word — signs found meaning as its structuring units.

We can think of signs, in fact, as the atomic units of aboutness, underpinning meaning the way atoms underpin matter. Like atoms, on close scrutiny signs reveal no simple presence but a dynamic substructure, and detailing this has involved sign theorists, or semioticians, in reasoning sometimes as arcane as particle physics, if without so much math. This book, however, will not be an exercise in semiotic theory. It will not take up Continental semiotics at all, chiefly because this body of theory has been slanted steeply toward humanist parochialism. Its various representatives, including Ferdinand de Saussure, Roman Jakobson, Algirdas Greimas, and Roland Barthes, concerned themselves especially with human language and discourse, and they made striking advances in an area that guided many disciplines in the humanities and social sciences through the late twentieth century. My orientation in semiotic theory calls instead for a widened approach that will engage the transspeciesism described earlier, and I find it in Peirce's semiotics, following in a line of other recent neo-Peirceans such as Deacon and Paul Kockelman (Deacon 1997, 2012a; Kockelman 2013). Peirce certainly can bandy semiotic arcana with the best of Continental thinkers, though in fact he developed his full-fledged semiotic system as much as a century before them, and many Continental theorists of generations after Saussure, including Derrida and Gilles Deleuze, recognized the distinctive priority of his approach (Derrida 1976; Deleuze and Guattari 1987). We do not need to don Peirce's full panoply, however, but only arm ourselves with a limited range of concepts that can intersect with today's animal and evolutionary studies. First in importance among these, we will see, is the interpretant.

If Peirce has become so important today that is because his semiotics offers an embryonic model of cognitive behavior that can link up with more recent thinking. The model features a *process involving organisms and their perceived environments* in which the subatomic structure of a sign is realized. Cognition creating signs is necessarily not only embodied but also embedded in the interaction of an organism and its surroundings — or "en-niched," as we might say once we have taken up the fundamental evolutionary process of niche construction (see Section 6).

The Peircean semiotic process in this way anticipated today's embedded models of cognition, extending from neural networks out to environments (Clark 2008), and the embeddedness of Peircean signs shifts the question of meaning away from internalist models of cognition and points it beyond the brain. In cognitive studies and philosophy of mind through the late twentieth century, this position swam against prevailing currents. The foremost model of cognition was a computational one, which likened the brain to a computer running symbols through programmed algorithms. The symbols were usually taken to be mental representations (there's that difficult word again), contents somehow lodged in the brain. Computationalists were always aware, of course, that perceptions of the external world had to play a role in the nature of the representations, but their focus was more on the manipulation of representations within the brain than on the space between brain and world. Recent connectionist alternatives to computationalism, which involve dynamic systems theory applied to the massively parallel circuitry of the brain, pay more attention to things outside the brain because the systems they envision can be opened to the external world more easily than systems that start from the premise of a separate, discrete representation of that world in mental tokens crunched by neuronal algorithms (van Gelder 1990). Nevertheless, most connectionist models remain brain-centered because the difficulties of modeling them are almost intractable even without trying to take account of external issues. Researchers tend to focus their efforts more on the changing states of dynamic neural systems and questions of how aboutness arises in these than on the relation of all this to body and environment.

Peircean semiotics does not obviate in one swoop the difficulties of modeling cognition or the "hard problem" of consciousness related to it. I make no such reckless claim. But it does offer a model that can accommodate, more readily than others, behaviors witnessed in nonhuman animals, and it focuses our attention on related capacities that are not evenly distributed in the biosphere. It opens out on a transspecies perspective and the richness of animal studies today, enabling us to make distinctions among the relations of various animals to their niches and to locate the borderline between meaningful and meaningless biotic complexity.

The entries that follow are arranged in ontological order, building from the bottom up an evolved world in which the transmission of information came to be enlisted and adjusted in several novel and distinct processes. *Information* and its *mediation* were *a priori* conditions on which this world was founded, but they did not remain unaltered as it developed, instead being partially reshaped by new possibilities and dynamics. *Sign* and *interpretant* are the chief tokens of the new possibilities. Why speak of newness or novelty? Because the evolved nature of this world brings the corollary that the ontological order of presentation is also chronological or historical. There was information in the world before there could be signs. This history also describes a *hierarchic* world, wherein later forms are dependent on and nested under earlier ones — or supervene on them, to use another philosophers' term. It also describes, finally, a *compounding* world, wherein earlier phenomena are not displaced by later ones but persist alongside them and as their infrastructure; signs could not exist in a world without information.

The world today, then, is more complex than it once was in respect to these phenomena, and evolutionary processes have made it so. This does not imply that earlier evolved behaviors did not manifest magnificent complexity, a complexity we can picture, for example, in the awesome predator *Anomalocaris* finning its way through Cambrian seas half a billion years ago (Gould 1990), or in the diverse cephalopods, ancestors of today's squids and octopuses, that ruled the seas for a hundred million years after that (Staaf 2017; Godfrey-Smith 2017). And it does not imply that behaviors sponsored today only on the basis of older phenomena — behaviors of a present-day ant or bee in its "superorganism" community, for instance (Hölldobler and Wilson 2008) — are not likewise magnificent. It does imply that the kinds of complexity exhibited and the modes of its construction distinguish animals operating within the sphere of meaning from animals and other organisms outside it.

THREE

Keywords

Information

Today a fault line crosses the conceptual landscape concerning information, one that rumbled its way to the surface in the information theory of the 1940s. In everyday usage of the word, information is associated with content, with "facts provided or learned about something or someone," as the dictionary on my hard drive puts it. But the dictionary offers a second meaning, less quotidian and less transparent: "What is conveyed or represented by a particular arrangement or sequence of things." This meaning hints at operational specificities and restrictions in a way the first does not. A particular arrangement or sequence must be bounded or limited, and to be informational, so must be the set of things it can comprise and arrange. An unending sequence built from an infinite array of tokens could not convey information. In the same way, an unending sequence of words spelled in an infinite alphabet could not carry content or provide "facts about something or someone," so these restrictions are at work also in the first definition of information, even though they are hidden. The two restrictions, limiting the number of discrete elements available for sequences and the length of the sequences, together describe a situation of *discrete combinatoriality*.

But note this difference: the sequences of the second definition are not tied to the content that is essential to the first. The initial challenge in constructing an idea of information broad enough to see its full extent is to abandon the idea that it is inevitably connected with content. Understanding meaning as a part, kind, or subset of information requires understanding also that information can be without meaning.

This difference — information conveying meaning vs. information without it — marks a division in modern conceptions of information, and the two kinds have been variously named. On the meaningful side, information has been called "intentional," referring to a philosophical notion of intentionality, that is, the quality of a state of mind that directs it at something or gives it an aboutness. It has been called "semantic" information, using the term with which linguists and logicians refer to meaning. And often it has been called simply "information," without qualification, reflecting the vernacular, commonplace scope of the dictionary's first definition. On the other side, the side without meaning, information has been named using terms corresponding to these. Plain "information" is countered by "Shannon information," after Claude Shannon, the early architect of information theory. The counterpart of semantic information is "syntactic" information, referring to linguists' syntax, the arrangement of words and phrases in language. Though this word captures well the arranged aspect of our second definition, it is not wholly satisfactory, since syntax for linguists is in no simple way separated from meaning and sense. The counterpart to intentional information, finally, carries implications more essential to the divide and to the conception of information at stake in this book; it is "causal" information. Why is causality basic to information?

When Shannon set about constructing his "Mathematical Theory of Communication" at Bell Labs in the 1940s, refining and extending the ideas of earlier Bell scientists, the problem he took on was to define formally the nature and conditions of a system "reproducing at one point either exactly or approximately a message selected at another point." He was quick to note that the "meaning" and "semantic aspects" of a message were "irrelevant to the engineering problem," which involved instead questions about source and transmitter, information channel connecting them, and receiver and destination (Shannon and Weaver 1949, p. 31). Information, in this view, is the correlation of a state at one place or moment (the source) with another state at another place or moment (the destination), a correlation brought about by a causal connection between the two. The generality of this statement allows us to see that there are myriad possibilities for the specific nature of the causality at work in the correlation. But, like any meaning that might be conveyed, these specificities are irrelevant to the engineering problem. What is crucial is the fact of causal correlation.

Decades after Shannon, philosopher Jerry Fodor provided a thumbnail definition of information as "reliable causal covariance" (1990, p. 93). This conveys the causal correlation basic to information, but it also introduces a separate idea, reliability, which is not an aspect of information itself but of its measurement or quantification, the main issue Shannon worked through in his paper. The quantity of information in any system is a measure of the uncertainty of a transmitted sequence being received as it was sent. The greater the uncertainty, the greater the information in a reliable transmission.

Calculating this uncertainty depends on both the length of the sequence and the number of choices among tokens available at each new unit in it. We can easily understand that the probability of identity between the states of transmitter and receiver is smaller for a message composed of thirty units selected from fifty possibilities than for a message of two units selected from two possibilities. Put in more abstract terms, the measure of fidelity between source and destination is a measure of the distance of their correlation from maximum randomness. So the positive measure of information is also a negative measure of entropy, with maximum entropy occurring where there is no information transmitted, no correlation of state brought about between source and destination. In all this, the quantity of information does not depend on any meaning that might be transmitted.

We could offer, then, another pair of terms for the two kinds of information: anti-entropic information vs. meaningful information.

But this would cloud an important fact that is hidden also in all the other pairs, that all meaningful information is anti-entropic, just like information without meaning. It always features causal covariance by virtue of its redirecting of a mind according to its content, however reliably it might do so. Meaningful or semantic information is a subset of Shannon or causal information, information carrying meaning a subset of information in general, whatever pair of terms we choose. The fault line we started with turns out to be a nested hierarchy or a Venn diagram with the set of information-bearing meaning completely inside the set of informational interactions all told.

This suggests also that some recent attempts to reduce all information to a phenomenon of the human mind — to consider it an "epistemic" phenomenon only (DeDeo 2017) — are misleading. No doubt quantifying information, theorizing it, and analyzing systems of its transmission are all uniquely human epistemic activities. No doubt (more broadly) any meaningful information is also a purely epistemic phenomenon, although it will be one of the aims of this book to show that this episteme extends far beyond human minds. Causal information in general, however, is an ontological fact extending far beyond minds, and arguably even beyond the living systems that support them.

To avoid slipping into the parochialism of many humanist accounts of meaning, we need a view of meaning as transhuman in extent. And to avoid semantic universalism, on the other hand, we need a view that generalizes causal information beyond any kind of meaning or mind. How far might this lead? Shannon's causal correlation (Fodor's causal covariance) suggests the answer: very far indeed, beyond living systems and even beyond whole biospheres. For our purposes, we can follow information to the edges of our biosphere and speak of biotic information, a ubiquitous property of the open thermodynamic systems exemplified by all living things (see Part II). The dictionary we started with seems to acknowledge the ubiquity of information to life in the illustrative phrase it supplies along with its second definition of information: "What is conveyed or represented by a particular arrangement or sequence of things: *genetically transmitted information.*"

A stance of universalism is justified regarding information but not regarding meaning. All living things are miracles of information processing, but their biotic information is not, in the vast majority of its operations, about anything. It is more fundamental than this, a sheer causality that makes things happen, from the molecular level on up through organismal and ecosystemic levels. It is the processual emergence of correlated difference itself in the complex systems at work at all these levels — what maintains them as systems or organized hierarchies of components and what opens them to their surroundings and the other systems (likewise open) there. The fact that aboutness and meaning sometimes arise from these networks of causality is a historical accident introduced to this openness.

Mediation

After half a century following docilely in the train of media, mediation has lately lifted its head and asserted its power and range.

Media studies at least since Walter Benjamin, and decisively since Walter Ong and Marshall McLuhan, have focused on media technologies and their distinctive impacts on human users. In this discourse, the term mediation has named the specificities of those impacts, the workings of technical apparatuses embedded in broader systems of thought, discourse, affect, habitus, and institutions. For such a system, at once technical, cultural, social, and perceptual, media theorists and in particular film theorists often adopt the term *dispositif* (French: apparatus) or dispositive from Michel Foucault's studies of institutional apparatuses (Casetti 2015, chap. 3). Mediation then refers to the processes at work in a specific dispositive. The term *mediation* also finds an important place in a separate intellectual tradition, Marxist cultural critique. Here the concept, descended from the *Vermittlung* of Hegel's dialectic of mind and phenomena, then refracted by Marx through the analysis of political economy, is tied to relations between capital and other aspects of society and, especially in twentieth-century critique, to the dynamics of ideology (Williams 1976).

In these usages the concept of mediation furthers particular disciplinary agendas, but four recent interventions by John Guillory, Richard Grusin, John Durham Peters, and Paul Kockelman outline its broader scope. Guillory makes the important point that the idea of mediation was not habitually linked to the nature of any specific medium until the twentieth century, and certainly not to the dispositives of particular technical media. In the nineteenth century processual emphases predominated in the use of the term. Guillory's two cases in point are Hegel's *Vermittlung* and Peirce's theory of signs, "the first full-scale theory of a specifically *semiotic* mediation." Guillory appreciates the novelty of Peirce's move, "violently displacing traditional philosophical questions into the domain of the semiotic," and understands that the displacement represents also a complicating of representation of any kind, exposing the "hidden complexity" of its structure. Representation is a mediated process, and the important Peircean term for this is *interpretant* (Guillory 2010, pp. 344–46).

In his theory of "radical mediation," Grusin duplicates Guillory's turn to Peirce and emphasizes Peirce's processualism. What makes Grusin's mediation radical (and in his view gives it a paradoxical *immediacy* in experience) is its immanent, generative force: It is "the process, action, or event that generates or provides the conditions for the emergence of subjects and objects, for the individuation of entities in the world" (Grusin 2015, p. 129). Hegelian ontological *Vermittlung*, Marxian mediation, Peircean semiosis, media studies, technics studies, and communication theory are all subsumed in a mediation viewed as the mobile, transformative, transducing flux that creates — more than merely connecting — things and experiences. Such mediation, we can see, is poststructuralist in its emphasis on the openness of all seemingly closed structures and entities. For Grusin it extends far beyond human experience, and he allies it with a range of concepts including Deleuze and Guattari's assemblage theory, the exploded agency of Bruno Latour's actor-network theory, Karen Barad's "agential realism," in which "intra-actions" of matter generate structure, and the panpsychism of some object-oriented ontologists. This network of associations is more evoked than systematically pursued by Grusin, and we will not follow him quite so far, but we can take away from his argument especially the broad-front expansion of the process and consequences of mediation into the nonhuman biotic world.

Peters's impact on the question of mediation is indirect, since for him it is the concept of media itself that needs expansion. His "philosophy of elemental media" inverts a century of anthropocentric media theory in a paratactic survey of water-, fire-, and weather-media that finally circles back to the humans trying to make sense of it all: to their writing, their information networks, their anthropocentric limitations of vision and concept. Peters offers an environmental media theory — or, as we might say, a *niche-constructive* media theory — in which his three elements of fire, air, and water are infrastructures fashioned into "media for certain species in certain ways with certain techniques" (Peters 2015, p. 49). *Homo sapiens* is one of innumerable niche-making, media-building species, embedded like them in elemental media. For Peters this warrants expanding the range not only of media but of meaning as well, and here again Peirce's semiotics is exemplary: "To posit media of nature is to deny the human monopoly of meaning. Media can be rich in semiotic stuff without being the sole property of humans; indeed, as Peirce understood it, semiotics was the study of all signifying activity, from protoplasm to God." This implies a broader interpretation of Peirce's main thinking on semiotics than I will advocate in what follows. Indeed, extending the reach of signifying activity so far, and with it the reach of meaning, Peters outstrips even the semantic universalists: "Nature abounds in meaning, most of which we have no idea how to read. . . . There is an exquisite pattern in DNA and the neurons of sea slugs, in photons and the red shift, in the bonds of the carbon atom and the fortuitously odd behavior of water" (pp. 380–81).

In the work of Grusin and Peters we sense the danger of expanding the concepts of media and mediation, and with them meaning, to the point where important distinctions are blurred or lost from view entirely. In order to reap the benefit of their thought we will want to rein in this expansiveness in a way that restores the distinctions, particularly in regard to the relation of mediation and information. What follows is a quick sketch of the different kinds of mediation such a delimited approach allows.

Mediation entails, first, something or some process in the middle. (Peters quotes a sixteenth-century maxim of Aristotelian bent: "*Nihil agit in distans nisi per aliquid medium* — nothing acts at a distance except through some middle thing" [p. 47].) This mediate thing can be likened to the channel of Shannon's information, and this is more than a happenstance resemblance. It signals the necessity, both for mediation and for information transmission, of some kind of medium, support, support system, or, in Peters's favored word, infrastructure. The infrastructural requirement is clear enough to see, even in the most extended views of information. We could, for example, move beyond the biosphere and say that one billiard ball hitting another, and the second one hitting a third, involves a transmission of information; the situation surely fulfills the basic condition of causal covariance, after all. Even here the information transmission requires an infrastructure, manifest at least in the table surface, cushions, atmosphere surrounding them, and gravitational field. In biological systems, with their ubiquitous information processing, the infrastructures reveal themselves even to casual scrutiny, though a more careful look always reveals daunting complexities that arise from the openness of these systems. The production of proteins by messenger RNA provides an example at the molecular level. mRNA carries a genetic sequence transcribed onto it from DNA that corresponds to a sequence of amino acids, the building blocks of proteins. But it cannot construct proteins from the coded amino acids without an infrastructural array including transport proteins that move the mRNA out of the nucleus, huge protein structures that form pores in the membrane bounding the nucleus, many other protein molecules regulating these processes, other RNA molecules (transfer RNA) carrying appropriate amino acids to the mRNA, and ribosomes — themselves structured of a third sort of RNA (ribosomal RNA) and dozens more proteins — where the attachment of one amino acid to the next is catalyzed (the "reading" and "translation" of the mRNA's sequence). From the molecular level on up to the level of whole ecosystems, infrastructures ground all biotic information transmission.

If biotic information always involves an infrastructure, it is also always mediated; conversely, it is hard to think of any biotic mediation that is not information transmission. This amounts almost to an equation of the two, except that in thinking about mediation we focus on the process of the transmission. The infrastructure of information viewed in the processual light takes us back to the media theorists' dispositive, only in a widened sense, encompassing dispositives throughout the biosphere. Mediation is the traversal of the dispositive, the crossing of the intermediaries that compose the infrastructure. In biotic systems it is the very opening of life.

Here then is another way of thinking about biotic information: *the mediation of states across material infrastructures connecting open entities ranging from molecules through metabolic systems all the way to organisms, populations of organisms, and the cultures that some few of these have produced.* The breadth of this statement is meant to encompass not only biotic information without meaning but also all the meaningful information that arises from and only from it.

The mediation as described so far appears to be a linear process; think of the billiards example, with one ball struck by another and striking a third: $A \to B \to C$. Most examples of mediation in complex systems, and especially in the wetware dispositives of biotic systems, add something onto this simple scheme. They involve also nonlinear (or reciprocal, circular, or cyclic) relations among their components, as when the third billiard ball rebounds off a cushion to strike the first: $A \to B \to C \to A$. Here the mediation closes a loop, feeding back to effect change in an earlier stage or an earlier link in the apparatus. Such feedback designs are fundamental in the evolution of life (see Sections 5-7).

53

A third type of mediation differs qualitatively from these, since it adds a new level to the relations in the dispositive:

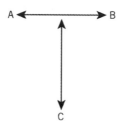

Figure 3.1

Here *C* relates not to *A* or *B* alone but to the *relation* between them. Perhaps *C* creates this relation; perhaps it facilitates it or alters it somehow (as in, for example, a chemical catalyst). Whatever its effect, a relation to a relation is at stake, so that we can speak not merely of relation but of *metarelation*. This kind of mediation shows multilevel, hierarchic, and recursive aspects, which become clear if we extend it and compare it to likewise extended linear or nonlinear (i.e., feed-back) relations, respectively, $A \to B \to C \to D \to E \to F$ and $A \to B \to C \to D \to E \to A$. Here is the extended diagram of relations to relations:

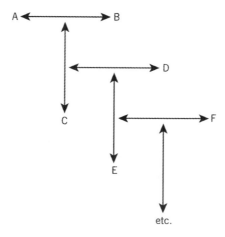

Figure 3.2

In this diagram, just as C stands in relation to the $A \leftrightarrow B$ relation, D stands in relation to the $C \leftrightarrow (A/B)$ relation, E in relation to $D \leftrightarrow [C/(A/B)]$, and so forth. A recursive chain of burgeoning metarelationality is formed.

This brings us to the last and most dramatic of our four expansions of the mediation concept, by anthropologist Paul Kockelman, who has emerged over the last twenty years as a true theorist of metarelationality, in a line including Marx, Saussure, and — most importantly for Kockelman and for us — Peirce. Kockelman describes "semiotic ontologies" involving agents in networks of sign-making, but, following Peirce, he sees this semiosis as metarelational through and through (Kockelman 2013, 2015). It involves a three-part interrelation of sign, object, and interpretant, each mediating the relation of the other two such that, in Figure 3.1, $A–B$ is mediated by C, $B–C$ by A, and $C–A$ by B. For this we need a new diagram, with double arrows indicating relations and single arrows indicating relations to relations:

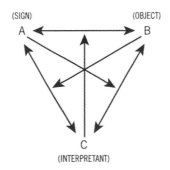

Figure 3.3

In Kockelman's view, agency is the organism's sensation of situations and instigation of actions arising from or enabled by this metarelational process. His organismal agent is embedded in its environment — opened out to it — by virtue of these relations, so deeply embedded and opened, indeed, that he writes at times of an organism/environment unit, the "envorganism," rather than either separately. Here he comes close, as Peters also does, to the niche construction I will describe in Part II, a foremost dynamic in extended evolutionary models. Kockelman understands all life-forms as "forms-of-mediation," and, since media are those things that mediate, he pronounces them, with characteristic punning, also "media-in-formation" (2013, pp. 38-39).

Kockelman extends his metarelational mediation very far, so that it finally encompasses even natural selection. This involves him in an all-out personification of a "selecting" environmental "agent" with which he is palpably and rightly uncomfortable (see pp. 29-31). In doing this he extends the dynamic of signification, and with it meaning, to lengths that approach semantic universalism. For him, for example, every exploitation of an environmental affordance by any organism can be seen as a semiotic process (p. 48). We will not follow him so far, as we did not follow Grusin or Peters; instead I will describe a narrower range of semiotic ontologies and offer reasons for this delimitation. In the sign itself Kockelman emphasizes the symmetry of metarelational mediation, as we can see in the three equivalent arrows of Figure 3.3. In the Peircean model I will pursue, instead, the mediation by the interpretant of the sign/object relation takes pride of place in the semiotic process. This is because it points us, more clearly than the other two vectors of mediation, toward perceptual processes involved in sign-making.

Despite these qualifications, we can take away from Kockelman three valuable messages. First, however far meaning might reach in the biosphere, we must understand it as the result of mediation of a metarelational nature. Second, the Peircean interpretant will be central to this understanding. Finally, failure to take account of the sign's metarelationality — a failure that has taken several forms, including emphasis on the structure of the sign rather than the process of its making and neglect of its tripartite nature in favor of the sign/object dyad alone — will limit our understanding of meaning. This failure was, Kockelman concludes, the "fatal flaw" of many twentieth-century theories of meaning (p. 51).

Sign

The sign, the minimal unit of aboutness, entails a substructure of mediation and metarelational processes. So, to follow the atomic simile one step farther before leaving it behind, the sign is not the atom of Rutherford and Bohr, a tiny, stable solar system with electrons revolving in their neat orbits around the nucleus. In the Peircean view it is more like the dynamic atom of Heisenberg and Schrödinger, conditioned by but not wholly determinate in the complex interrelations of its components.

Here is one of the best known of Peirce's many definitions of a sign:

> A sign or *representamen* is something which stands to somebody for something in some respect or capacity. It addresses somebody, that is creates in the mind of that person an equivalent sign, or perhaps a more developed sign. That sign which it creates I call the *interpretant* of the first sign. The sign stands for something, its *object*. It stands for that object, not in every respect, but in reference to a sort of idea, which I have sometimes called the *ground* of the representamen. (Peirce 1955, p. 99)

The passage names the three parts of the sign introduced in Figure 3.3 and also a fourth element, the ground of the sign. This last element is not part of the sign itself, but instead the conditions of cognition it depends upon — an infrastructure, as we can say with Peters. For Peirce the ground involved a fundamental tendency of thinking organisms to enter into habits and to form then "habitual connections" between percepts (Sheriff 1994). I will return in Part II to a post-Peircean view of this ground and its extent in the biosphere, for now noting only that it is not limited to human mental processes, as Peirce's vocabulary here seems to imply ("somebody," "person").

Peirce's definition characterizes the three elements of the sign carefully, if telegraphically. There is the sign itself, also called representamen, a Peircean term useful for its connotation of the representing in the sign of something else, but which we will leave aside. Peirce also called this element the "sign vehicle" — once again a useful term, since it distinguishes between the sign vehicle as an element in the three-part structure of signification and the sign as the product of the interaction of all three.

The sign vehicle stands for something, its object, to something, the perceiver of the sign. The first of these relations, to the object, is not a relation of one-to-one fullness, that is, a relation in which the sign captures every aspect of the object. This would be an impossible signification, rendering the sign identical to the object and collapsing the whole process. Re-presentation depends on relay and deferral; it cannot be identity. Instead, the sign isolates certain aspects of the object to form its relation to it (it stands for the object "not in every respect"). Thus a deer-crossing sign on a highway portrays the silhouette of a leaping deer — the shape of a particular body in motion — to form its relation to the object. Other aspects of the deer (its furriness, its true color, its real size, and so forth) and other aspects of the sign (its yellow and black colors, its diamond shape) are not relevant to the signification. In this way the sign/object relation is always *aspectual* and *partial* (see also, for this point, Atkin 2013). In Section 12 we will see that this fact carries important ramifications for our understanding of the capacities that bring about signification in animals.

The aspects of the object involved in any instance of signification are mutually determined by the sign vehicle and the object. A sign vehicle necessarily constrains the ways an object can relate to it. Peirce described three general types of this constraint: sign vehicles that rely on quality or felt experience, those that rely on things in the world in their eventful, processual singularities, and those that rely on generalizations or "laws" about these events. The most famous of several intersecting typologies in Peirce's analysis of the sign-process arises from this distinction of the kinds of aspects a sign might isolate in representing an object. A sign that relates to its object through a qualitative likeness he called an *icon* (the deer sign is an example). A sign that relates to its object by indicating or pointing to it, or in a relation of proximity or contiguity, or in a causal connection is an *index* (smoke to fire, for example). And a sign related to its object through conventions, rules, and laws is a *symbol*. This last is arguably (and to say so will indeed cause arguments) a kind of connection only humans make in the world today, along with, perhaps, some nonhuman animals they have trained. Its foremost example is a word in human language.

Just as the sign vehicle constrains what aspects of the object can enter into signification, so the object constrains the sign vehicle. In the example of the deer-crossing sign, a silhouette of a humped shell can play no role, though it is part of the turtle-crossing signs one sees now and then on country roads. Every object of a sign shapes the aspectual possibilities that can enter into its signification, and we can see that this shaping will be an important dimension of the environmental niche of any semiotic organism, since the constraints of object on sign vehicle will be patterned according to what is available to perception in a niche — its perceivable affordances. Here we see again the necessary openness of semiosis to the organism/niche relation, and we veer close to Kockelman's envorganism and Peters's elemental media.

Where does the interpretant enter into this semiotic system of mutual constraint? Peirce's sign vehicle stands in a relation to its object such that it "addresses" a perceiver in a way that elicits a second sign. This is the organismal response — to the sign vehicle, in constraining connection to its object, and on some ground of perception — that creates the interpretant. The mutuality here is a complex one, both because it depends on the evolved capacities of any semiotic organism and because it features the metarelationality of the sign in a dual way, with the interpretant determining *and* determined by the relation of sign vehicle to object.

The interpretant was a critical revelation of Peirce's semiotics and one of the foundations of his expansive thought, and I will examine it in more detail; but there is one general implication of its role that needs introducing here. Peirce specifies that the interpretant is itself another sign (Peirce 1955, p. 100). The content or meaning of a sign as a whole is present to a perceiving mind in the form of a sign. (This is not the mental representation of the computational cognitivists, but an element in the ongoing process of signification itself.) The second sign, which is the interpretant, must have its own interpretant, like any sign. This second interpretant arises in response to the relation of the first interpretant to the initial sign/object relation, so that, in relation to the second, the first is a sign vehicle taking as its object the initial relation of sign and object *as a whole*. Here is the deepest role of metarelationality within the semiotic process, brought about by the interpretant. The three parts of the first sign triangle collapse into two (sign-vehicle and object), with the third element of the new triangle supplied by the new interpretant. What we see emerging in this process is the pattern of Figure 3.2, with D the second interpretant, relating to the initial interpretant C in its relation to A/B. Following the chain further, E is the interpretant of D in its relation to $C/(A/B)$, and so forth. This is the unending sign chain for which Peirce was most well known in late twentieth-century semiotics, especially after Umberto Eco focused on it to adjust and open earlier structuralisms (Eco 1976).

This enchained dependency of one sign on another reveals again the systemic openness basic to the semiotic process. All sign-making is recursive in the manner of Figure 3.2, and every tripartite sign structure (as pictured in Figure 3.3) is not independent or static but a link in the chain. The chain also shows sign-making to be hierarchical, since the metarelationality of the sign extends not only among the three parts of a sign (as in Kockelman's view; Figure 3.3) but all along the chain, with each new link nesting the triad before it as its own sign/object dyad. In Sections 12 and 13 I will argue that this hierarchic nature is related to another one among signs, which has to do with the typology of icon, index, and symbol, and results from the evolved capacities animals require in order to perceive signs. Recursive hierarchy and individual aspects related to wholes are two features of perception and cognition that are required for signs and meaning to emerge.

The enchainment of semiosis does not characterize all metarelationality, which can take the simpler form of mediation illustrated in Figure 3.1. In other words, semiotic enchainment is not an ontological *a priori* — not, in this sense, metaphysical — any more than are the signs that arise from it. Earth and life on it existed for billions of years before there were signs. Signs arise from a mutuality that an animal enters into in the context of its lived environment, a relation in which affordances take shape that constrain and conform to both the animal and its lifeways. There are no affordances except by virtue of the animal's exploitation of them, but also no lifeways except by virtue of the animal's fit with what an environment offers. If there is something here akin to Heidegger's *Zuhandenheit* or "readiness-to-hand," it requires a more expansive, far-wider-than-human vision than Heidegger brought to the question.

In this offering-and-exploiting according to processes arising from certain capacities, semiosis takes place. We can now redescribe the sign. It is something formed when an organism's perceptions enable not merely the exploitation of an affordance but a perceiving of affordance *through* another resource or entity — another affordance. Each constrains the other in ways that also constrain the percept as it can form the linkage for the semiotic animal. The animal at the same time constrains what of the interacting affordances can enter into the linkage.

Interpretant

The fullness of these relations shows Peirce's interpretant at work, in a metarelational elaboration of informational causality that only certain animals can bring about. A few more points need to be made concerning this all-important aspect of signification. The interpretant concept underwent a convoluted growth across forty years in Peirce's thought (for good accounts, see Bergman n.d. and Short 2007). Here we focus on a late snapshot.

The interpretant marks the appearance of metarelationality in the sign-process and its three interacting parts since it arises as a sign of the relation between the "initial" sign and object, where the scare-quotes remind us that there can be no *a priori* element among interpretant, sign, and object in the formation of the sign; all three are constituted together in mutual metarelation. Peirce described the interpretant as a product of these multiple mediations in 1907:

> A sign endeavors to represent, in part at least, an Object, which is therefore in a sense the cause, or determinant, of the sign. . . . But to say that it represents its Object implies that it affects a mind, and so affects it as, in some respect, to determine in that mind something that is mediately due to the Object. That determination of which the immediate cause, or determinant, is the Sign, and of which the mediate cause is the Object may be termed the *Interpretant*. (Peirce 1994, vol. 6, ¶347)

The interpretant is an effect (a "determination") brought about immediately by the sign vehicle and in mediated fashion by the object, both of which come into those roles by the determining action that is the interpretant. This metarelational mediation is what most needs explaining in semiosis as an organismal relation with the world because it concerns the organismal response that fulfills, provisionally closes, and makes significant the sign process; hence it is what most needs explaining about the emergence of meaning. The interpretant is a process in the relation of the sign-making organism to its environment, its niche, its *Umwelt*.

The determination of the interpretant is not, however, an interpretation on the part of the semiotic organism in which it approaches the objective world and with sovereign authority forges links where it will and assigns them meaning. This is a common misunderstanding, unfortunately encouraged by Peirce's term. There is in the interpretant no interpretation at all, in the usual sense of this word. Instead, the interpretant is a calling of the organism by the sign vehicle that shapes the vehicle's mutually constrained relation to the object. This is the "address" to the perceiver in Peirce's definition of the sign quoted before. In addition to his typologies of the three natures of the sign and of its three relations with the object, Peirce offered several tripartite typologies of interpretants, determined by the nature of this sign/object calling. From the intersections of all these "trichotomies" together emerged an overall, complex typology of signs for which Peirce's mature semiotics is well known.

The calling alone, however, does not adequately describe the interpretant. It is, as we saw, a sign created in response to the initial sign/object relation, and this second sign reflects the organism's reaction to the calling of the first, according to its evolved capacities to interact with its environment or niche. To understand the interpretant, we must embrace not only the calling but also this response, each an exerting of constraints on the other in respect to information from the world. Peirce sums up the capacities for response, in both of the preceding quotations, with the eminently humanistic word *mind*, but we will see in Part II that organismal responses to things in the world such as to make of them sign/object pairs reach far beyond what is parochially thought of as mind — far into the nonhuman animal world and the heightened cognitive powers of many species.

Because of these multiple mediations in the signifying process, the phrase *sign-making*, though a necessary shorthand, is something of a misnomer. The organism is brought to signification through the impact on it of aspects of its surroundings, which aspects are linked to one another through its evolved capacities as well as through their natures. The overdetermination of the dependency of each of these on the other — response dependent on call and call on response — carries us, once again, deep into the openness of an organism to its niche. Its native capacities enable it to enchain perceptions concerning things in the world so as to generate another sign, the interpretant. The objects, entities, and ideas taken into the sign triad are not innocent and outside the processual dynamism of semiosis but are constituted by the organism's structuring of affordances: constituted, in other words, by all its embedded interactions with its environment or niche.

So the interpretant is characterized by a flow of reciprocal impacts, a complex of feedback relations among the parts of the sign, and the special, metarelational complexity of these interactions. If the sign as a whole is a perceiving of an affordance through another affordance, the interpretant is the process in which an animal is called to and also activates that perceiving-through. Where could sovereign interpretation enter into such a rich amalgam? Only as a short circuit, halting the dynamism, openness, and recursion of signification as a lived experience in an ongoing lifeway, or as a purely humanist and so deracinated view of the metarelations at work. To understand the full implications of Peirce's work, we need to imagine the interpretant call-and-response as a bird or elephant manifests it. And we need to see why many other complex animals, not to mention plants and microbes, do not manifest it at all.

Signal (vs. sign)

The conditions of possibility for signification are far more exigent and less widespread than those for information alone. These conditions set a high bar for meaning, carving out for it a small corner in the vast realm of information in the biosphere. There is one more common term that needs to be constelled with this information/sign pair. Signal is a word easily confused with sign and sometimes used interchangeably with it. The customary usage of the word, however, shows that it connotes a kind of thing less specific than a sign, opening a wide window on the causality that operates in information.

Among ethologists, signal is a ubiquitous term connoting a huge array of interactions between one animal and another, "any act or structure which influences the behavior of other organisms (receivers)." Animal communication is predicated on the presence of signals, "a process involving signaling between a sender and receiver, resulting in a perceptual response in the receiver having extracted information from the signal" (Stevens 2013, p. 73). In this informal usage, a signal is the basic unit in a model of information transmission like Shannon's, with transmitter, message, and receiver. And, like Shannon's model, it is neutral in regard to the distinction between causal and semantic information and so neutral in regard to the question of meaning. All animal interaction that brings about altered behavior, of any sort and across any medium — sonic, visual, chemical, electric, tactile — can be described as signals sent and received. As ethologists realize, this broad application opens the term out to an even broader one, involving interorganismal interactions across the whole biosphere, plants and microbes as well as animals (see Bradbury and Vehrencamp 2011).

Beyond ethology, signal appears in all manner of biological discussion, not as a real term of art but as something less technical and more colloquial. Here are several examples:

On the regulation and transport within the cell of building-block molecules involved in its structuring and growth: *"Specific recognition signals incorporated into the structures of proteins and nucleic acids route these molecules to their proper cellular compartments. Receptors recognize these signals and guide each molecule to its compartment."* (Pollard and Earnshaw 2008, p. 8)

On how "core processes," basic genetic and metabolic networks strongly conserved in evolution, can exert diverse regulatory controls across the lives of multicellular organisms: *"Weak linkage refers to specific biochemical features of information systems in biology, where signals of low information content evoke complex, preprogrammed responses from the core process."* (Kirschner and Gerhart 2005, p. 265).

On the establishment of environmental features in the organism/environment relation rather than independently of the organism: *"Organisms determine by their biology the actual physical nature of signals from the outside. They transduce one physical signal into a quite different one, and it is the result of the transduction that is perceived by the organism's functions as an environmental variable."* (Lewontin 2000, p. 63)

On the nature and varieties of biological information transmission: *"The [sender-receiver] model can be applied to signaling between organisms and within them, and also to cases where the boundaries of the organisms are not clear. Signaling can occur across space and time."* (Godfrey-Smith 2014, p. 149)

In these passages, taken respectively from a cell biology textbook, a study of the emergence and stabilizing in evolution of regulatory mechanisms, an essay on the nature and structuring of life-forms' environments, and a survey of the philosophy of biology, the idea of the signal is applied to a wide range of biological phenomena occurring at radically different scales. These include intracellular molecular reactions, genetic and epigenetic mechanisms effecting correlation and change across the tissues and organs of multicellular organisms, and the interface of an organism with its environment. Only in the last of these examples does the interorganismal communication of ethologists appear, and even here its appearance asserts the equivalence of inter- and intraorganismal applications of the term. Signals, in such usage, can be the ubiquitous relations by which processes of metabolism are switched on and off and homeostatic balances within and beyond the organism are maintained or altered. And they can initiate *signal transduction pathways*, complex biochemical cascades set in motion in cells by external stimuli, from receptors in cell membranes all the way to nuclear genetic materials, where input is transduced from one kind of physical phenomenon (light, for example) into another (electrochemical impulses in an optic nerve).

In ethology and beyond it, then, the deployment of the term signal generally reflects the legacy of information theory in biological thinking, but it is the broad category of causal information, not semantic information, that the term is tied to. The uses of the word here do not reveal anything about signification because signals, like Shannon information in general, need not carry any meaning. They are changes of state in one organism, or one part of an organism, or one aspect of an environment that trigger correlated changes in another. In such usage, signals need not involve any perception on the part of an organism. You and I do not perceive when a particular level of blood sugar signals the pancreas's production of insulin.

Signals characteristically (though not always) operate to activate a programmed, inevitable response. This can be true even when a signal is a percept. A honeybee sensing particular chemicals on nother honeybee has two options, forage or not. This is not a choice or decision on the bee's part but a preprogrammed dichotomy, and, if concentrations of the chemicals are sufficient in certain contexts, the foraging response is inescapable. The context-dependency here does not open for bees a realm of decision making, but instead involves the conditions within the hive, all of them created by and creating additional systemic chemical balances, additional signals.

These examples suggest the connection, very frequent in the biological literature, of the term signal with phenomena involving a threshold in chemical concentrations, mechanical stimuli, or energy levels. Signals in this usage are a kind of information functioning to effect (reliably) Fodor's causal covariance in relation to border-levels of one kind or another. A sign, in contrast, is not a threshold phenomenon but a *linkage* process — a multiple, metarelational kind of linkage, as we have seen. Signals need have none of the complex structure involved in any sign; in particular, no metarelations are required for them to function. They are inevitable parts of a lifeworld, but signs are not.

PART TWO

The Abstract Machines of Evolution

How can we locate the creation of meaning in the history of life?

What are its mechanisms, and how are these related to

the most general mechanisms at work in evolution?

FOUR

The Evolutionary *Mise en Abyme*

Ever since Darwin, evolution has toyed with the scientific imagination. It is as plain as day, as solid as any conceptualization of the cosmos modern science has yielded, and study of its consequences has long since transformed it from a theory or hypothesis into a finely woven tapestry of known facts about earthly life. Nevertheless, its mechanisms are intangible, trajectories immanent in its processes. The entities it has produced appear to constitute chiseled natural categories, but these blur on closer inspection, their edges eroded by unceasing change at every temporal level. Many terms we adopt to describe these mechanisms and entities suggest a fixity that proves provisional or nonexistent. Other terms are analogies or metaphors, at once as helpful and misleading as such things can be. Constant motion, labile categories, and not fully determinate concepts are the stuff of evolutionary science. Here are some instances.

Natural selection, Darwin's epochal insight, remains foundational today. Nothing in the history of life makes sense except in its light, to redirect a famous phrase of evolutionist Theodosius Dobzhansky. But it is inaptly named, since it entails no selecting or choice. Darwin devised his term, in *The Origin of Species*, to exploit the explanatory power of its analogy to artificial selection, the deliberate choosing by human breeders of traits to perpetuate and enhance (Darwin 2003, chap. 1). Already in 1860, just one year after the publication of *The Origin of Species*, he worried about his choice of the name in a letter to geologist Charles Lyell, but the die was cast (Darwin 1888, p. 2:318).

Selective pressure is an expression commonly used to name the action of natural selection on organisms, affecting their ability to reproduce according to their relations with their environments. But it is only as a metaphor that we can think of selection as acting in any way, since it entails no physical mechanism at all. It is better described as an algorithm guiding interactions of matter and energy (Campbell 1974, 1983; Dennett 1995; Odling-Smee, Laland, and Feldman 2003) or, as we will describe it here, as an abstract machine (Deleuze and Guattari 1986) — a schema of conditions in which energies and entities meet and fall into a process. Another term, *fitness*, referring to the way organisms are accommodated to their environments, has been fundamental in qualitative descriptions of natural selection since Darwin's day, and by the mid-twentieth century it was formalized in quantitative analyses of the effects of selection. But success in producing progeny can come about for any number of happenstance reasons, not only because of advantages thought to be selected, so it's hard to link fitness transparently to selection. In order to be tractable, meanwhile, the equations of quantitative analysis must be stripped of most of the multidimensional complexity of replication in the living world. What traits or features of organisms constitute fitness is difficult to say.

Adaptation might well be the most fundamental term after natural selection itself, but it reveals puzzles and misdirection — a little like Marx's commodity, a concept developed about the same time. A common definition would be a trait selected for the advantage it confers on the organism bearing it — dinosaur feathers selected for thermoregulation, for example. But the timing here is out of joint. Given that selection is the differential survival of organisms on the basis of traits they (already) possess, how could those feathers be selected for that function before they were in place and functioning? How can any trait be selected *for* anything, when its emergence is an *a priori* condition for its advantage to be felt? One strategy in the face of this difficulty is to talk of *exaptations*, traits evolved for one purpose and later co-opted for another — those heat-regulating dinosaur feathers later proving useful for flight. This seems a coherent idea, and we can imagine small, feathered dinosaurs profiting from the ability to glide short distances long before full flight was enabled, thus enhancing their chances of surviving and reproducing. But the concept of exaptations only passes the buck. The assertion, frequent these days, that all adaptations must be exaptations only circles back to the initial adaptive puzzle. For with each exaptation we are still faced with the impossibility of an original selection-for.

The puzzles of adaptation engulf even the *traits* supposed to be adaptive and their *functions*. Terrence Deacon points out that natural selection is "mechanism-neutral," that is, any physical properties might come to be functional in the midst of shifting environments or selective contexts. An adaptation, then, "is the realization of a set of constraints on candidate mechanisms, and so long as those constraints are maintained, other features are arbitrary. But this means that with every adaptation, there are innumerable other arbitrary properties brought into play." Think once more of dinosaur feathers, whose heat-dissipating structural properties — in relation to one set of constraints — turned out to carry with them aerodynamic properties also, which butted up against another set of constraints. "Any of these incidental properties," Deacon continues, "may at any point themselves become substrates for selection, and thus functional" (2012b, pp. 423–24). An adaptive function in this view is a virtual thing, a dynamic process arising from a play of virtual constraints upon physical properties. The adaptive process destabilizes traits also, since they are always aggregates in its dynamic, composed of multiple properties with multiple functional possibilities in relation to shifting environments. The deepest, hardest to eradicate flaw in adaptationist thinking may not be its imagined, temporal causal chain where there is none, but instead its abiding belief in the discrete, readily identified trait-for-a-purpose.

As with traits, so with the clumped types we witness in the living world; all of them are processual and dynamic. Natural selection results in a diversification of types or kinds that are difficult to comprehend due to the very dynamic that produces them. Phyla, genera, and other taxa result from complex histories, and they are impossible to define in necessary and sufficient simplicity. "Only that which has no history can be defined," Nietzsche wrote in *The Genealogy of Morals* (2:13), and, though his maxim concerned a certain kind of cultural, not biological, process, he was on to something of more general import. Biological kinds, as outcomes of history, are as difficult to make conform to general principles or laws as cultural kinds: the Reformation, the French Revolution, or — Nietzsche's case in point — kinds of punishment practiced in human societies (Wagner and Tomlinson 2021). Evolutionist Kim Sterelny gives us a less aphoristic version of Nietzsche's maxim when he writes that "historical processes destroy evidence about their own causes" (Sterelny 2003, p. 3), but to focus on lost evidence would be to simplify Sterelny's message. The difficulties stem already from the processualism he names.

Thus *species*, the fundamental entities prompting Darwin's intervention, are notoriously difficult to isolate, and those brave enough to try point to as many as two dozen competing criteria for defining them, most of which involve processes that might have brought about diversification. "Groups of organisms that reproduce together," the most common of these, is insufficient in many instances, from bacteria to many plants and to the nonreproducing insects in colonies (Godfrey-Smith 2009). *Genes*, naming another crucial biological kind, are slippery things that have been rendered less rather than more determinate by the abundant new knowledge about their operation gained over the last several decades. They are, we now know, spliced, diced, and recombined at every turn; and their expression in phenotypic traits is always biased, canalized, regulated, and switched on and off in complex ways by other genes, nongenetic molecules, and environmental stimuli. These processes have led many to view a gene not as a stretch of nucleotides in a DNA molecule but as a complex involving a diverse set of molecules drawn together into a specific metabolic function and responsive to external stimuli, which further unfocuses our view of a gene as a discernible entity. Some even

74

argue that genes are epistemological entities rather than physical ones, pragmatically defined in differing ways to assist one research project or another (Waters 2004; Dupré 2012).

We finally seem to reach solid ground with the *organisms* already mentioned; what could be more straightforwardly bounded and defined than a paramecium, ant, elephant, or oak tree? But colonial organisms such as jellyfish render the terrain treacherous again, not to mention superorganisms such as the colonies formed by ants. And what about the now-famous estimate that there are many more foreign microbe cells in the human gut than there are human cells in the whole body? What, in this context, does it mean to speak of a "human" cell? We live our lives colonized from within (by symbionts) and without (by parasites and other invaders), and we cannot easily define the border of even so palpable an organism as a person. This means that the very distinction of organism and its surroundings must be carefully considered, as it begins to soft-focus into something like Kockelman's envorganism, encountered before in Section 3.

Perhaps in the end the entity that admits the most universal clarity is the *cell*, the basis of all earthly living things. "*Omnis cellula ex cellula*," "all cells come from cells," aphorizes the nineteenth-century realization that, among its other effects, once and for all excluded spontaneous generation from the halls of science. But mustn't there have been at least one spontaneous transition from inorganic systems to something that we can consider alive, from non-life to life, non-cell to cell, marking the beginning of earthly life?

It is in the midst of these ramifying puzzles that evolutionary biologists, historians of life and its forms, pursue their work. Those of theoretical bent use generalizations they understand to fit many but not all cases and models they know to be applicable only in restricted ways, like all models of complex realities, including those of history. Stephen Jay Gould's lampooning of the just-so stories of adaptationists — they reminded him and coauthor Richard Lewontin of Dr. Pangloss from Voltaire's *Candide*, with his faith that everything was arranged to turn out for the best (Gould and Lewontin 1979) — is right enough, but also tendentious, since *all* evolutionists, Gould and Lewontin included, are makers of provisional narratives. They share this with historians of other kinds, and in every case questions of nuance, marshaled evidence, and attentiveness to theoretical conundrums are important in separating real work from fantasy, allowing us to distinguish a compelling narrative from a crocodile stretching an elephant's stub nose into a trunk. Today the most ground-gaining narratives turn away from static definitions of the terms here reviewed toward something more pliable and mobile: attempts to capture the dynamism and historical motion that characterize all evolving phenomena (see Dupré 2012; Nicholson and Dupré 2018; Bueno, Chen, and Fagan 2018). We aim, in other words, at *process* more than definition, and in this emphasis we follow Darwin, whose immense, enduring achievement was to identify a process on which our explanations of evolutionary change can depend, if not rest.

What follows starts from first principles to describe generalized models of several processes basic to evolution. These are abstract machines, patterns that set in motion particular processes: given these conditions bumping up against each other, this process involving materials and energies will begin. The abstraction of the machine, the distancing of the schemata from the substrates in which they generate processes, is necessitated by the multiple realizations they allow, as Deacon points out. And the abstraction has the advantage of always reemphasizing the provisional nature of the kinds named earlier, that is, genes, cells, organisms, species, traits, and adaptations.

FIVE

Darwin's Abstract Machine:

Natural Selection

Life itself is hardest of all to define, a fact well known to philosophers of biology and authors of biology textbooks everywhere. Still, life-forms show certain regularities we can think of as quasi-universals, at least. I'll call them universals by way of shorthand, while remembering that, in matters of life and evolution, one should never say "always."

Life-forms, in the first place, reproduce, generating new versions of themselves. These inherit much from their parents, resembling them in many features and traits, but, given the violable, non absolute nature of the molecular processes forming them, they also vary from their parents and from one another. Varied inheritance in living things was the starting point for Darwin's reasoning in *The Origin of Species*, and the exploitation of it by pigeon fanciers, cattle breeders, and flower cultivators explains why he began his book, unexpectedly for first-time readers today, with the artificial, human selection mentioned earlier (Darwin 2003, chap. 1). Varied inheritance remains a cornerstone for evolutionary thought, whatever other building blocks it might use, and it probably is a condition for living systems anywhere they might occur.

A second universal: life-forms, even the simplest, even those like viruses that cannot reproduce independently and that stand at a borderline of organization that casts doubt on whether they are alive at all, always involve differing systems joined in interaction or opening out to one another. In the case of a virus particle there are two: genetic material and a protein casing for it. Even in the case of the hypothetical, not-yet-living *protocells* that most theories today accord a place at the origin of earthly life, there are distinct systems: a lipid membrane sac separating inside from outside and a resulting concentration of molecules inside different from that outside and likely to catalyze the production of one another in the process called "autocatalysis" (Maynard Smith and Szathmáry 1995, Kauffman 1993, 1995). Multiple systematization pertains in all the life-forms descended from such protocells, and it means that they are internally differentiated and hierarchized into part and whole, at least rudimentarily. Evolution occurs in respect to the differentiated structures of organisms, while an undifferentiated anti-organism (a block of salt crystal, for example) cannot evolve, at least not in the same sense as life-forms do.

Structural differentiation, moreover, implies differential change in the parts, the hierarchized systems and subsystems, making up an organism (Eldredge et al. 2016). Variation from generation to generation never takes the form of equal change in all the systems making up an organism; at a molecular level such a notion could make no sense. Instead organisms pass changes in their features differentially from one version of themselves to another, with one or more parts or subsystems of it varying in reproduction while others do not. There might of course be complex correlations between the changing system and others. Such correlated change was a point of emphasis already for Darwin, though he knew no mechanisms to explain it. But evolution works in general with organisms varying in partial and system-specific ways.

The permeability of the boundaries of organisms I exemplified before with human bodies and their symbionts is also fundamental to living things — another universal. Life-forms are not isolated systems, like a mixture of sugar and water in a closed jar, moving according to the second law of thermodynamics toward equilibrium and maximum entropy. Instead they are systems open to their surroundings (Bertalanffy 1969). They subsist on exchanges with what is outside them, taking in energy or material proxies that yield energy and channeling these through loops of chemical, metabolic processes to build subsystems, drive further processes to produce more energy, and reproduce subsystems and whole organisms. Life-forms inhabit an energy space far from equilibrium with their environments, which requires mechanisms to stabilize it of a complexity that has been characterized as hovering at the edge of chaos (Kauffman 1993). To stay in such a space, achieving and maintaining the mechanisms of structural differentiation, requires marshaling energy and the work it drives.

The external energy and matter exploited by organisms are usefully named *affordances*, a term coined by perceptual psychologist James J. Gibson to name a "complementarity" of an animal and its environment, judged in terms of what the environment furnishes (Gibson 1979, p. 127). Affordances are features from outside an organism that have surmounted a threshold to become useful for it in its maintenance of its far-from-equilibrium state. They gauge the manifold modes of openness of any organism to its environment. Innumerable environmental features, both biotic and abiotic, can serve as affordances: sunlight and CO_2 for photosynthetic plants, O_2 for deer and mountain lions, water for lions, deer, and plants, deer for lions, plants for deer, and so forth. Ecologists divide organisms into *autotrophs* — "self-feeders," such as photosynthetic plants — and *heterotrophs* — "other-feeders," such as deer and lions. Even the autotrophs engage in affordance-relations with other organisms; subterranean fungi, for example, facilitate exchanges of inorganic nutrients between the soil and the roots of plants.

Today the word affordance is often paired with another word, *constraint,* in an effort to be more specific about what the environment offers or denies the organism, but there is no real difference between offering and denying. Gibson himself thought of affordances as environmental provisions "either for good or ill." The affordance/constraint pair is a complementary one: A constraint is the negative measure of an affordance, and an affordance is never unlimited, always constraining. Photon energy in sunlight might seem a limitless affordance for photosynthetic organisms, but at night or on a cloudy day it reveals its limits and shows itself as constraint. Constraints are the same as affordances not only ecologically but also in the light of evolution. Organisms do not subsist hypothetically, that is, in relation to features of the external world that might be helpful if they were present; instead they are shaped in relation to what is there. What is not there is irrelevant, and, though changing what is there by removing some feature of the environment might be deemed constraining, it can bring about organismal change to accommodate what is *still* there, still an affordance.

Organisms join with one another in defining the affordance/constraint complementarity. The overstory of a forest canopy monopolizes sunlight, limiting the photonic resources available to plants growing on the forest floor below. We might speak of tall plants *constraining* the growth of small ones, but we could speak equally well of the environmental *affordances* for the growth of the small plants, which include the tall ones — and have resulted, in many such species, in the reduction of "shade avoidance" mechanisms found in other plants (Pigliucci and Kaplan 2006, p. 147). For the plants below, the tall trees are at once affordance and constraint.

We already encountered the idea of constraint in Deacon's view of adaptations. Following in a line of systems theorists and biologists including Ludwig von Bertalanffy, Ilya Prigogine, Humberto Maturana, Francisco Varela, and Stuart Kauffman, Deacon uses the term in a thermodynamic sense: A constraint is a restriction or confinement of a system within certain bounds, impeding either its maintenance of its non-equilibrated state or its move toward equilibrium (2012b, chaps. 6–7). In either case, work against constraints forms the basis for the creation of order and organization, and so, just as constraint is complementary to Gibson's affordance, it is complementary also to organization or structure. Deacon sees constraints as either extrinsic, arising in the open relation of organism to environment, or intrinsic, within the organism, as for example in the action of DNA to determine specific molecular interactions within the cell. However, given the formation of DNA and its constraints in the cell by natural selection and other evolutionary processes, and given the continuous regulation of gene expression along pathways reaching out to the environment, the extrinsic/intrinsic distinction disappears, finally, in the broadest view of the thermodynamically open organism.

This broad view explains why all affordances are limited. Just as organisms are structured in relation to environments, the affordances of those environments are defined in relation to organismal capacities and behaviors. To adapt an avian example offered by Richard Lewontin: The tarmac road outside my home is an environmental affordance for seagulls that drop clams and oysters onto it to break them open; for the egrets fishing in the lagoon next to it, it is not. The lagoon, hunting ground for both shellfish and fish, forms an affordance for both, but for each kind of bird — this is Lewontin's point — its particular *reach* out to certain features around it, but not to others, defines its environment (Lewontin 2000). Egret and gull share the same space but live in different environments. Since the effective reach of any organism toward the surrounding resources is always limited, affordances in any environment are also limited. The limitation reflects not only how a resource comes to be an affordance in its use by an organism, but also how an environment comes into being in relation to the organisms in it.

The finitude of affordances joins in a working relation with varied inheritance and leads us to the heart of Darwin's *Origin of Species* and to the first, foremost abstract machine of evolution. It is the abstract design that results in natural selection, what William Calvin referred to in 1987 as the "Darwin Machine." Differences among organisms, however subtle, define (however subtly) differences in their reach out to resources and in the resulting creation of affordances. Given the limited quantity of any affordance, one variant's increased capacity to exploit it will give it an advantage over another variant not so endowed. Here is Darwin's machine:

Condition 1: Inherited features vary

Condition 2: Affordances are limited

Condition 3: Different organisms differ in their use of affordances

$\downarrow$

Process: Natural selection

Darwin's own reasoning on this point was famously influenced by the population principles of Thomas Malthus (Darwin 2003, chap. 3; Malthus 2008). But whereas Malthus was interested in the wax and wane of human populations in relation to human productivity, Darwin's focus on inheritance with variation enabled him to see broader vistas. Natural selection is the inevitable outcome of reproduction of organisms occurring at greater rates than can be sustained by their affordance-creating reach. It results from the operation of an abstract machine that, as long as its conditions are met, is self-perpetuating.

Let's pause to understand the extraordinary kind of machine at stake here. Abstract machines are schemata that define the emergence of self-organizing structural states in dynamic systems. They generalize interplays of matter and energy that can be realized in countless physical situations. Manuel de Landa, in a typology of abstract machines extrapolated from Deleuze and Guattari (1987), arranged them in order of the increasing complexity of their schemata (de Landa 1997). Most basic is the *attractor* machine, manifested in the tendency of many dynamic systems toward an emergent stable order — a whirlpool vortex in a stream, for example, or convection cells in energized fluids of all sorts. A *sorting* machine, next, introduces hierarchy into its emergent stable state — a rushing stream sorting lighter stones and particles over heavier ones and thus forming strata in its bed, for example. Such sedimentation contrasts with a more complex structural hierarchy in which the dynamics of organization enables some sorted elements to form containers, networks, or meshworks for others. De Landa's example of this third machine, the *meshwork machine*, is granite formation, in which the different thresholds of crystallization of diverse ingredients in magma result in late-forming crystals interlocking in organized ways with earlier ones.

Darwin's machine comes fourth in line in de Landa's typology, and it requires a crucial additional element in the dynamic system, overlaid on attractor, sorting, and meshwork dynamics: replication with variation. The combination of these designs results in what de Landa memorably calls a "blind probe head" (p. 264), the effect of a search space in which combinatorial possibilities or forms are explored; in other words, natural selection. This probe head schema is arguably the fundamental characteristic of any dynamic system that can be said to evolve. A block of salt represents in its crystalline structure either a sorting or meshwork machine, depending on foreign elements other than NaCl that might be in it, but it has no potential to evolve in its formation.

The presence of varied inheritance in the abstract Darwin machine results in processes organized in a certain way that Malthus already described, and that we today refer to with several terms: nonlinearity, reciprocal or circular causality, and *feedback*. Natural selection is a process in which the environment determined by a population of organisms feeds back onto it through environment/phenotype interactions. This feedback reaches through its effects on the phenotype to the genomic resources of the population, altering its nature in future populations. The alterations, expressed in altered phenotypes, are markers of evolution. The complicity of varied inheritance in this feedback process makes it fundamentally different from most homeostatic feedback systems. A thermostat is a feedback mechanism that enables switching between two finite states of a system (furnace on or off); James Watt's flywheel governor was another feedback mechanism maintaining a more-or-less constant level of operation in his steam engine; and the software driving your laptop embodies countless algorithms with feedback loops designed into them. But, unless you're building AI or artificial life programs

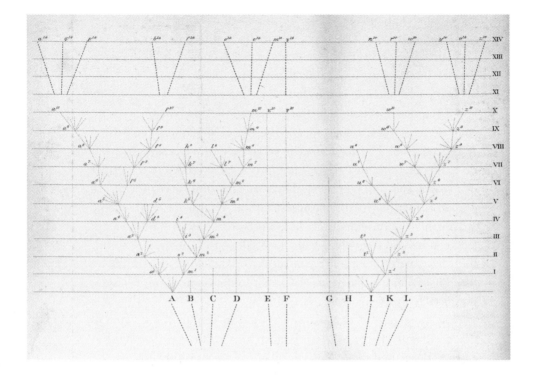

Figure 5.1. Darwin's diagram of diversification, extinction, and genus formation, from *The Origin of Species*. Earlier species *A–L* at the bottom; each horizontal line represents the passage of *x* generations; branching trees above *A* and *I* represent the emergence of new varieties, subspecies, or species, including many extinct variants; descendant species at top form two groups branching from *A* and *I*, which might represent new genera.

on your computer, these algorithms do not involve lineages or varied inheritance, with its constant input of novel conditions shifting the feedback dynamic as a whole. Not one of them evolves.

Though the idea of a search space lay in a post-Turing future, Darwin intuited that populations of organisms, as they move through successive generations passing along features that sometimes vary, in effect search a space of shifting relations with their environments that confer on one variant or another greater or lesser reproductive success. His visualizations of species, subspecies, and varieties exploring search spaces usually were tied to their real-world environments. This remains true even when he describes hypothetical cases, thought experiments such as three populations of sheep with varying traits, two living better on broad upland and lowland ranges, respectively, the third better on a narrow range in between (Darwin 2003, p. 676). The variant in the middle, he suggested, competing for its range with variants on both sides, would dwindle to extinction — the victim, as we might say, of blind probe heads encroaching from two different directions.

In one brilliant gesture Darwin visualized a de Landa–like *virtual* search space that sums up all the major consequences he discovered in his analysis of natural selection. His astonishing diagram in *The Origin of Species* of diversification and extinction among species ranks high among the most powerful graphic formulations modern science has offered (Figure 5.1). It captures many things: the branching across time of life-forms into bushy or treelike patterns; the descent of various species from a common ancestor (the converging dotted lines at the bottom, presumed to meet in a single taxon below the reach of the diagram); the formation, across many thousands or millions of generations, of new genera from a single species (the two topmost groups stemming from *A* and *I*); the possibility of "living fossils," life-forms persisting with little change across long periods (*F*); and the probable extinction of intermediate forms, like those sheep caught in the middle (see the many intermediate, dead-end branchings above *A* and *I*). All this powerfully represents phenotypic spaces of organismal possibility being searched and explored.

A Post-Darwinian Abstract Machine:

Niche Construction

A second machine follows from the first. It too was glimpsed by Darwin, though he had other business to attend to, other positions to advocate, and did not give it pride of place. It is not much in evidence in his diagram. Later, in the second quarter of the twentieth century, when Darwin's natural selection was joined to Mendelian genetics and the statistical analysis of populations in the *modern evolutionary synthesis*, the second machine was veiled, often disappearing completely from view. In recent years, however, it has resurfaced to become one of the mainstays of today's *extended evolutionary synthesis*.

The second machine, like the first, arises from the organismal determination of environments and the limitation of affordances. But it emphasizes an additional consequence of this limitation: The exploitation of affordances by organisms must alter the balance of resources from which they are created, whether matter or energy. Organisms thus change the horizon of possible environments that other organisms or future generations in their own lineage can create. No organism is so light-footed in its lifeways that it does not have this effect — that it moves through its world without changing it. No organism leaves no footprint.

In the most general way this is an inevitable consequence of the thermodynamic openness to the external world of all life-forms

(Odling-Smee, Laland, and Feldman 2003, p. 168). Were we to turn the tables and think for a moment of ecosystems as organisms — they are complex open systems of open systems, after all — we could view the living organisms in their midst as *their* affordances. (We cannot keep the tables turned for long in this game, since in many other ways organisms are not like ecosystems, especially in reproducing versions of themselves and in the precision of the metabolic homeostasis by which they maintain their far-from-equilibrated energy states.) Sometimes the changes in environments can be immense and global, even when they originate in tiny organisms. Starting as long as three billion years ago, the first photosynthetic microorganisms gradually built up free oxygen in Earth's atmosphere. As this "Great Oxygenation Event" took hold, the anaerobic microorganisms that had once thrived mostly died off, in what paleontologists consider the earliest mass extinction. The stage was set for the last billion years' proliferation of photosynthetic and aerobic organisms.

While the first abstract machine emphasizes the alteration of kinds of organisms by their environments, the second emphasizes the two-way relation between them, with environments altering organisms and organisms altering environments. Environmental alterations of organisms work at many different timescales, from short, quick changes relying on greater or lesser plasticity of individual organisms all the way to natural selection altering the most stable features of organisms and populations of them. Alteration in the other direction, that is, organisms' alteration of their environments, seems simpler at first glance:

Condition 1: Resources in any environment are limited

Condition 2: All organisms use resources (as affordances)

↓

Process: Alteration of environments by organisms

The alteration of environments does not involve anything like the selection that over time alters lineages of organisms. This is because environments do not involve the reproduction with variation of lineages, a fundamental condition for Darwin's machine. Inevitable change characterizes environments, just as it does organisms, but not evolution through selection.

The consequence of organisms' alterations of their environments has moved toward the center of recent evolutionary thought under the name *niche construction*. As living things build their environments, shaping their ecosystems in combination with the other organisms around them and with abiotic forces, they construct their niches. As I noted, Darwin perceived something close to niche construction, particularly in the hypothetical instances he described of the coevolution with one another of separate varieties or species; his example of different kinds of bees and clovers coevolving is one of many memorable thought experiments in *The Origin of Species* (Darwin 2003, p. 611). But niche construction was not front and center in evolutionary thinking in the mid-twentieth century. Evolutionists' formal analyses of fitness tended to leave the environment in the background, as an element presumed to be more or less static. One of the most engaging manifestations of this thinking, the *fitness landscape*, shows the trend (see Wright 1932; Kauffman 1993; Gavrilets 2004; Pigliucci and Kaplan 2006; McGhee 2007). In its typical form (see Figure 6.1), this is a three-dimensional graph in which the horizontal x and y axes at the base plot increasing frequencies of two phenotypic traits or genetic variants (or alleles): for example, average body weight in different ancient hominin species measured along x and average cranial capacity of the species measured along y, or the differing average sizes of varieties of ammonites on x and a gradation of features of their spiral shells on y. The z axis measures the fitness of the various combinations of these. There is no place in such a graph — no fourth-dimensional axis available — for a shifting environment.

Even when twentieth-century evolutionists took on the changes organisms bring about in their environments, they often viewed these

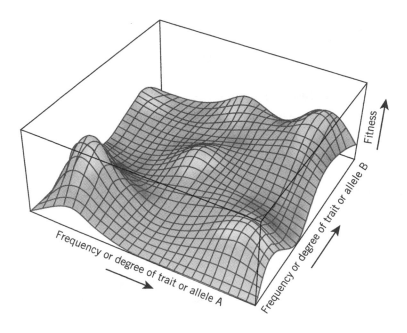

Figure 6.1. Graph of a hypothetical fitness landscape.

in limited terms, as a one-way impact of organisms on environments. Richard Dawkins, for example, famous for his gene-centered views and the idea of the "selfish" gene, in 1982 enlarged his views to take in organisms' environmental alterations — beaver dams, termite mounds, and the like. But he regarded the changes as gene-controlled "extended phenotypes" of the organisms involved. The straight-line causality from his selfish genes on out was maintained, only lengthened beyond the organism itself to take in its environment (Dawkins 1982).

Niche construction theory calls for something more, something that enriches the second abstract machine and endows it with its full heuristic power (Odling-Smee, Laland, and Feldman 2003). Natural selection acting on organisms and their genomes, on the one hand, and niche construction by organisms, on the other, are not independent dynamics but must always be interdependent. We see this if we set in motion the process defined in the second machine, as in a moving picture. What comes into view is a relation between organisms and their environments that admits of no stasis in either one. Environments are defined by the affordances organisms in them exploit, and these in some fashion affect natural selection and determine organismal survival. But the ongoing alteration of the environment by organisms continually changes the affordances, changing the dynamics of selection acting on organisms and so changing, ultimately, the genomes of populations and species.

The second abstract machine in this way defines, like the first, a relation of circular or reciprocal causality, a feedback loop (Odling-Smee, Laland, and Feldman 2003; Tomlinson 2018). But the feedback cycle of natural selection was mainly an organismal affair, with the interaction of one population with its environment feeding back to alter a later population in the same lineage, all in the face of affordances conceived to be static or neutral. Niche construction inserts the environment into the loop. This is hard to reduce to three-dimensional visual analogies such as fitness landscapes, but it can be schematically represented in the circular path of a flow chart from organisms to environments and back again:

$$\text{Genome} \longrightarrow \begin{array}{c}\text{Phenotypic}\\\text{trait}\end{array} \longrightarrow \begin{array}{c}\text{Environmental}\\\text{alteration}\end{array} \longrightarrow \begin{array}{c}\text{Phenotypic}\\\text{trait}\end{array} \longrightarrow \text{Genome}$$

$$\uparrow$$

$$\begin{array}{c}\text{Natural}\\\text{selection}\end{array}$$

Here, in a simplification we will complicate later, we see genetic information determining a trait and the trait altering the organism's environment in the course of its niche construction. This in turn changes the selection on the same trait or others, in the same organism or in others in the ecosystem. And this altered selective dynamic finally results, across many generations offering new variation, in an altered genome in the affected organisms. It is foreshortening, then, to stop at the idea advanced earlier that environments are defined by organisms. Instead, both organisms and their environments dance in ever-shifting relations with one another. They are mutually determined in an ongoing cycle, extending temporally from momentary fluxes to millions of years' worth of change.

The feedback loop of niche construction extends continuously between organisms and environments. Across evolutionary timespans, altered organisms always present altered relations to affordances, and these adjustments must alter their environments in new and different ways. Beaver dams are not merely an extended beaver phenotype directed by genes (Dawkins's model), and we cannot gauge the fitness they confer in an environment taken to be stable (the fitness landscape model). Instead, they change the hydrology of whole areas and regions across many generations, with far-reaching consequences for the environments of countless organisms, including the beavers. These consequences shift the dynamics of selection on the organisms involved. Across their evolution, castorid ancestors of modern beavers became more and more adapted to aquatic lifeways and dam-making (Rybczynski 2007), in a feedback involving a growing prevalence of aquatic niches as their environments, a prevalence created by the behaviors of the castorids themselves.

Niche constructionists consider the environmental side of this loop an externalized inheritance system, alongside the internal inheritance systems of organisms (Odling-Smee, Laland, and Feldman 2003; for a summary, see Tomlinson 2018). They speak of an "ecological inheritance" bequeathed by earlier organisms to later ones. This inheritance usually changes environments in incremental ways. Ecologies can be transformed abruptly and catastrophically by meteors, volcanoes, and other forces, but niches are most often defined through slow shifts and generational transmission. We can illustrate this gradualism by imagining again the Great Oxygenation Event. It is not enough to think of moment 1, when there were no photosynthetic algae, and moment 2, when the algae appeared, anaerobic microorganisms retreated to what were now "extreme" environments, and everything was changed. We need instead to imagine a biological calculus of infinitesimal changes, filling in the gross picture of moments 1 and 2 with as many as two billion years' worth of organismal and environmental shifting between them. From an anaerobic global atmosphere to an aerobic one, one microbe at a time.

Here is a more complex view of the second abstract machine, incorporating the spinning feedback dynamic of organismal and ecological inheritances:

Condition 1: Resources in any environment are limited

Condition 2: All organisms use resources (as affordances)

↓

Process 1a: Organisms alter their environments

Process 1b: Altered environments alter selection on organisms

Process 1c: Organisms altered through selection alter environments in new ways

Process 1d: Altered environments alter selection in new ways

Etc.

Here again, as in Darwin's machine, there is no tangible mechanism as such, only a set of conditions binding organisms and environments in looped, nonlinear interdependencies. This machine, however, takes in the whole of Darwin's machine and adds to it an additional array of feedback loops by which organisms and environments mutually determine one another. And importantly, for my purposes, it describes, more explicitly than the first machine, forms of mediation, which we saw in Part I to be the traversal of apparatuses characterizing all information transmission. In the flowchart on p. 91 we see phenotype mediating between genotype and environment, environment mediating between different phenotypes (or the same phenotype at one moment and a later one), and genome mediating between one phenotype/environment relation and another.

In a deep historical *durée* spanning billions of years, both the niche construction machine and the natural selection machine were spontaneous emergences. From the moment when a biosphere was formed, when autopoietic entities coalesced and the first organisms took shape, all the conditions of the two machines fell into place. When they did so, the complexifying, diversifying, entangling, and reciprocal processes they drive began their looping operation. (We can remain agnostic as to how the originary moment came about, though the theories today are multiple and richly developed; for three recent views see Luisi 2016; Kauffman 2019; Deamer 2020.) In their abstractness the machines are vastly extendable as dynamics in the evolution and diversification of life. We might be justified in thinking that their processes have been and are at work wherever life has formed in the cosmos — that they describe, in other words, the conditions of a "universal biology" (Sterelny and Griffiths 1999, chap. 15). In the same way, we might be confident that those nonearthly biospheres would involve the other universals introduced before: an anti-entropic relation of open organism-systems to their environments, internally differentiated organisms, reproduction and inheritance of traits, and the inexact nature of that inheritance, with the resulting variation of traits.

It seems that these things might govern life wherever it appears, just as it must be based on physical laws or on the chemical nature of elements in the periodic table, with their different bonding valences and potentials to build complex molecules. Popular sci-fi depictions of aliens reveal the frailty of our imaginations in conceiving life-forms different from those on our planet. Green-skinned, big-eyed, bipedal, Area 51-style aliens are counterintuitions in our experience, but minimal, trivial, parochially humanoid ones. Have I enacted a similar frailty in extending across the universe the general conditions of life in these machines? Perhaps, but if so it is very hard to conceive alternatives. The conditions of the abstract machines seem to be foundations for evolution, so to imagine life-forms not featuring them would be to imagine a non-evolving life — something like our block of salt. In what biosphere could such a thing be alive?

SEVEN

Extended Evolution

and the Mediation Machine

One more feature that is immensely generalizable concerns the mechanism of inheritance with variation. It requires some sort of material code governing the transmission of features across generations of organisms, a code that can both found their similarities and vary within groups of like organisms, sponsoring their differences. Such a code has, as its own requirements, two features abstract enough to warrant our identifying its operation as part of a third abstract machine. It must use a determinate lexicon of distinct tokens, and it relies on a material infrastructure that can fix or stabilize them in finite sequences. Together these features define a situation of *discrete combinatoriality* — like a limited alphabet that can spell a large array of words. We met this combinatoriality earlier: It is characteristic of information in its broad, causal form, and indeed its two abstract requirements found an abstract machine of information.

In all earthly life the two conditions of finite lexicon and material infrastructure are met in RNA and DNA molecules. The structure of the latter was discovered in 1953, but the more general need for a discretely combinatorial, molecular "script-code" was put forward a decade earlier by physicist Erwin Schrödinger in his famous lecture series *What Is Life?* (Schrödinger 2012). There he described a large molecule arranged in an order more complex and irregular than the consistent, repeating, "periodic" structure of a crystal. This "aperiodic solid," Schrödinger saw, "endowed with sufficient resistivity to keep its order permanently," could be a "material structure that offers a variety of possible . . . arrangements, sufficiently large to embody a complicated system of 'determinations' within a small spatial boundary" (p. 61). The arrangements, it turned out, are orderings of four bases in DNA and RNA molecules. The code is built from sequences of these bases of varying lengths, and the infrastructure — the backbone of the double helix of DNA or the single strand, helical or not, of RNA — is formed from a sugar group and a phosphate group to which each base is attached. Together each base plus its backbone makes up one of the nucleotides in the DNA and RNA chains.

By the 1950s, a congruency was apparent to many between thoughts like Schrödinger's about the discretely combinatorial structures required for inheritance in organisms and Shannon's theory of information. Information theory joined explicitly with genetics and evolutionary theory. To view organisms as informational is first of all to understand that they materialize this abstract informational machine. To view them as inheriting features of their ancestors — the Darwinian starting point — is to see them as conveying informational sequences from one organism (a source or transmitter) to another (a receiver), however faithful or not the conveyance might be. It is to see all organisms as information processors.

And not trivial information processors. If information is quantifiable as the uncertainty that a given sequence of tokens will be received as it was transmitted, all organisms, even the simplest, are miracles of faithful information transmission in which hundreds of thousands (at a minimum), or millions, or billions of nucleotides in DNA molecules do their work from generation to generation with errors or alterations that are proportionately exceedingly rare. The work they do, moreover, concerns not only reproduction but also all the metabolic processes that maintain a cell. Correspondences of many sorts depend on them from the molecular level, where the production of proteins depends on the matching of messenger RNA codons (sequences of three nucleotides) to particular amino acids, all the way to the phenotypic level, where the code results in the consistent reproduction of traits from one generation to another. I'll complicate this simple picture; here the point is the awesome fidelity of the information transmission.

The informational correspondences embodied in the processes of metabolism are core aspects of all organisms' maintenance of their states far from maximum entropy. So one measure of the biological information coded in any cell involves the energetic distance it maintains from a nonliving equilibrium with its surroundings. Shannon's quantification of information, likewise, is a measure not only of distance from uncertainty, but also of distance from maximum entropy. Information has sometimes been regarded as a metaphor when applied to living things (see, for example, Godfrey-Smith 2014, chap. 9), but this congruency of information and anti-entropy suggests something else: that information conceived in Shannon's way cuts to the heart of the molecular mechanisms of inheritance and metabolism universal in living things, and that this is what we recognize when we think of them as information processors. It is only the term *information* that is metaphorical—and this only because we casually think of it in human, meaningful ways. There is nothing metaphorical about the combinatorics involved, whatever we call them. Looked at in this way, the emergence of information theory and machine computation in the middle of the twentieth century appears not as a set of technical and mathematical innovations later applied metaphorically to life-forms but as something more like the converse, innovations developed hand in glove with the growing understanding of the molecular coding implications of Mendelian and, later, molecular genetics. This is one message of Schrödinger's lectures.

The basic requirements of the code, as I've said, form the foundation of a third abstract machine. Beyond them, however, the flow of information in living things resists expression in any simple abstract schema, such as those laid out for the first two machines. This is true because of the complexity of the apparatus or dispositive. Even the chain of relations between the molecular structures of DNA and RNA and their proximate material effect, which is to produce the proteins necessary for cellular maintenance, growth, metabolism, and reproduction, entails many layers of mediation and a bewildering variety of macromolecules. The image often applied to this chain is one of a stream in which the genetic material is "upstream," the protein "downstream"; we can map the stream then as a long one, with many eddies between the two. Evolutionary-developmental biology ("evo-devo," for short) is the subfield most intent these days on understanding all the processes carrying us down the stream.

But the stream metaphor fails in an important regard. In a stream, whatever its eddies and back-currents, the water doesn't flow from downstream all the way back to the source. Yet the regulatory systems structuring protein production and involving DNA and RNA can do just that. As evo-devo biologists have taught us, the systems are looplike as well as chainlike, and molecular reactions far downstream — even outside the organism, in its environment — can circle all the way back to regulate the expression of "source" genetic material, governing its coordination with other genetic material and its production (or not) of the proteins corresponding to its codes. The molecules making up the systems can include, in addition to RNA molecules put to several different uses, proteins acting as *transcription factors*, which enhance or inhibit the expression of particular genes; hormones or pheromones, any of numerous kinds of chemicals that typically activate transcription factors; and "immediate early" genes, a first line of DNA sequences involved in producing transcription factors and so in regulating the expression of other, "late" genes.

We can trace here pathways allowing flexible response of the genome to changing circumstances which, in the case of many animals, can involve not only environmental changes but even the complexities of their moment-to-moment social interactions — in other words, a *modification of gene expression directly and quickly dependent on social change.* (I take up instances involving songbirds and bees in Part III.) This is not Lamarckism, and it should not even be thought of as neo-Lamarckism, since it involves a level of complication and granular understanding that leaves far behind any simple identification of traits altered in ontogeny and then passed to succeeding generations. Nevertheless, it calls for an extension of the feedback loops of niche construction to include momentary changes in gene expression (I will return to this), and it carries to the materio-energetic heart of the cell the kinds of reciprocal, nonlinear causality we found in the first two abstract machines.

Understanding these intricately mediated molecular systems has already brought about reconsideration of the nature of the openness of genomes to epigenetic and environmental alteration, showing the commonplace idea of genes as blueprints determining these processes to be not merely a simplification but a grave distortion. The systems have been assigned several names, depending on the particular slice of molecular machinery they describe, including the "character identity network" (ChIN) of Günter Wagner (2014), the "dynamical patterning module" (DPM) of Stewart Newman and Ramray Bhat (Newman and Bhat 2009; Newman 2010), the "core regulatory complex" (CoRC) of Detlev Arendt, Jacob Musser, and their coauthors (Arendt et al. 2016), and others (see in general Kirschner and Gerhart 2005). The networks, modules, and complexes referred to by these names are groups of molecules interlinked in the functioning of the systems, which can include all the varieties of molecules named earlier. By virtue of their networked functioning, the systems govern cell and tissue development in ways that open a wide molecular terrain, crisscrossed by two-way paths, between the nucleotide background and its phenotypic expression. They enrich and complicate the operation of the cellular dispositive, so to speak, with mediating elements and processes, and form in doing so nonlinear channels in the transmission of genetic information.

In a ChIN, for example (Wagner 2014), one gene sustains the expression of another and vice versa, and a whole set of mutually bound genes is required for the morphogenesis of the character or feature they (in a simpler view) "control." Genes in the network enhance the expression of genes outside the network that are additionally necessary for proper development of the character, and they suppress the expression of alternative genes and operation of other ChINs. To bring about these controls they join with complexes of nongenetic molecules, especially proteins. The mediating molecular networks form cohesive units whose functional linkages, once in place, create structural stabilities that resist changes that might be caused by mutating nucleotides; thus the whole system, including the characters it governs, enhances its own long-term conservation (Wagner 2014).

The effects of such regulatory networks are profound, at scales extending from single organisms to major evolutionary transitions, from moment-to-moment gene expression in a changing environment to stabilities lasting half a billion years. Their mediating functions can canalize or direct organismal development, and through such canalization, evo-devo researchers propose, the networks have shaped broad evolutionary phenomena such as the long, strong conservation of many homologies in animals and plants (ChINs) and the origin of metazoans, animals with differentiated tissue types (DPMs; see below). The robustness of such regulatory systems helps then to explain the persistence of core processes and structures in life-forms across very long timespans, for example, lipid membrane structuring, genetic transcriptional machinery, and bilateral body plans in animals. Such persistence is not well explained by natural selection favoring one allele over another, since millions of years of accumulating, selected change is a recipe not for the retention of traits but for their transformation. The mediation of stable molecular systems not directly subject to mutations offers a different kind of explanation for the persistence, and with it we enter a realm of directed or *facilitated variation*, in which the genomic outcomes of natural selection can be canalized through biasing processes involving the flow of information across dispositives of extragenomic molecules (Kirschner and Gerhart 2010).

A Cambrian example more than half a billion years old illustrates the sweeping impact facilitated variation can have. It advances a model for the rapid evolutionary appearance and radiation of trip-loblastic body plans of metazoans, that is, of animals with three distinct embryonic cell layers, from which differentiated tissues and organs are derived. All animals with distinct internal organs in the world today are triploblastic, from arthropods, worms, and mollusks to vertebrates. Newman and Bhat have proposed that this hugely consequential development was more than just selective in nature and was facilitated by a group of dynamical patterning mod-ules (Newman and Bhat 2009; Newman 2010). The DPMs involved were not apparatuses of genes themselves or even nucleotide chains, though behind them stood genetic "toolkits" (Carroll, Grenier, and Weatherbee 2005) akin to Wagner's ChINs, complexes of genes stably conserved and guiding development. Instead the DPMs comprised other molecules functioning in tandem with products of the toolkit genes to guide development. The genes and other molecules formed a mutually reinforcing feedback system that tapped physical proper-ties that had not been exploited in earlier organisms' ontogenies, thus introducing new morphogenetic possibilities.

Newman and Bhat propose that two DPMs in particular were of central importance in the advent of multicellularity, which they fostered by diverting certain proteins already present in single-celled ancestors to new functions inducing the adhesion of cells. The new, functional protein networks brought into play physical forces in the development of these organisms not relevant in their single-celled predecessors, processes exploiting viscoelastic chemicals that now operated "on the spatial scale of cell aggregates and tissues"—on an inter- rather than intracellular scale (Newman 2010, p. 286). From these novel functions came, first, the adhesion of cells to one another and, second, differential adhesion such that distinct layers of cells could form. And from these distinct layers evolved all future animals with differentiated internal organs. Stable regulatory systems such as Newman and Bhat's DPMs might thus facilitate major revisions in morphogenesis and then sustain them through long *durées* of natural selection.

I have already suggested, however, that they can also sponsor something like the opposite through the functioning of transcription factors and immediate early genes. The apparatuses' control of the expression of the genetic nucleotide sequences can operate even on a moment-to-moment basis, endowing organisms with a plastic flexibility to respond to their changeable environments. Compare this organismal plasticity with the plasticity of whole lineages created by natural selection: As natural selection enables lineages to search spaces of phenotypic variation — the spaces visualized in Darwin's branching trees of diversification (Figure 5.1) — so the regulatory mechanisms create within the ontogeny of single organisms what Marc Kirschner and John Gerhart call "exploratory processes." In these, mediation through extragenetic dispositives acts like "preprogrammed variation," processes that "maximize the amount of phenotypic variation for a given amount of genotypic variation" (Kirschner and Gerhart 2010, pp. 262–63; see also Kirschner and Gerhart 2005). The dispositives enable the building of molecular, metabolic networks that are available to be exploited according to environmental conditions, in turn enabling phenotypic plasticity. Examples of their products include microtubule structures in cytoplasm and axons extending from neurons. When they are exploited, integrated structures take shape — whole cytoskeletons within a cell or innervated muscle tissue — and phenotypic change results. When they are not, they do not affect the integrity of the regulatory complexes that stimulated their production. The neuronal changes enabled by such plastic mechanisms are particularly interesting in animals and may underlie memory-formation and learning processes (see Part III).

All these regulatory networks mediate the traits organisms present to their environments. This means, as I have suggested, that their effects on organisms' phenotypes are likewise effects on organisms' reshaping of their environments — on their niche construction. The resulting changes in niches might be a matter of resources used or byproducts produced more efficiently, with all the ramifying impact of these out through an ecosystem; these are much-discussed aspects of standard niche construction theory. But the changes might cut deeper than this and involve a resetting in which preexistent environmental resources not involved (or minimally or differently involved) in the prior lifeways of a lineage come to be incorporated in its revised lifeways. We can think of such a resetting as a change in the *affordance horizon* of the lineage. New resources rise over its horizon as new factors in its niche-constructive possibilities, and they do so because of the organism's phenotypic alteration or plasticity, at whatever timescale is relevant. Thus, in Newman and Bhat's Cambrian model, the new intercellular adhesive functions for certain proteins resulted in decisively new kinds of organisms able to exploit environmental resources in new ways: targeted predation of other organisms, for example, depending on distinct sensory and digestive organs. Even animal sociality can result from such shifting affordance horizons connected all the way to plastic molecular interactions, as we will see in Part III.

Such resetting of the affordance horizon is not a one-way affair, not a case merely of organismal adaptation. Instead it enters into the feedback cycles of niche construction, altering the selective terrain it presents to the organisms shaping it. The processes brought about by the mediation of molecular regulatory systems are all part of niche construction: the relations of molecular systems, cells, tissues, or organisms to their local ecologies. But this is niche construction with a difference, where systems integrated through *internal* feedback mutualities can loosen themselves from direct genetic control, altering in the process *external* feedback cycles with the environment at scales from momentary behavior to the evolutionary dynamics of selection. In this way these systems can change not only gradients in selection on organisms — this is the outcome generally ascribed to niche-constructive feedback cycles — but the range of possibilities of constructed niches in which selection can operate in the first place. Something more fundamental than an altered niche emerges: an altered kind of niche, with a shifted horizon of possible affordances. Conditions of possibility for niche alteration are altered, as well as the niche itself.

The semiotic machine I will describe in Sections 11–14, a fourth abstract machine of evolution, brings about just this kind of foundational alteration in affordance possibilities. The niches it enables some animals to construct are unavailable to other organisms; they are replete with meaning; and they enable the emergence of such features as advanced animal sociality, culture, and technics.

The complexity of mediation in most molecular regulatory networks warrants our assigning them a more emphatic adjective: they are not merely mediated but *hyper*mediated networks. This signals the multiple layers of their molecular dispositives, their many pathways and feedback loops. Meanwhile the efficacy of these networks at many different timescales, from those measuring evolutionary change in populations to moment-to-moment alterations of gene expression and niche affordance, warrants the idea of a *radicalized* niche construction, conceived as a single set of processes embracing everything from the interaction of whole populations of organisms on their environments across thousands or millions of years to the impact of an external stimulus on momentary genetic expression, leading quickly to an adjusted relation of one organism to its niche.

EIGHT

Toward the Question

of Meaning

Thoughtful biologists today — philosophers of life-forms — are everywhere involved in the unlearning of basic terms of their field (see Section 4). This movement, extending the twentieth-century evolutionary synthesis, represents a new understanding of what it means to historicize life. It involves new priorities and emphases: on change more than on defining or fixing entities, types, or kinds; on detailing hypermediated informational apparatuses at all scales; and, in the most general way, on process, that is, dynamics of development, organization, and reorganization. All these priorities also emphasize the constitutive openness of the systems involved. They are all aspects of the poststructuralist turn in evolutionary biology I described in the Preface.

Our increasing ability to characterize the historical processes involved in evolution, which we have described here on scales extending from the ecological (niche construction) to the molecular (evo-devo regulatory mechanisms), has set in motion all the terms that used to stand fast as foundations for evolutionism: genes, species, adaptations, and the rest. And it has enriched our understanding of the deepest, most solid foundation of all, natural selection, with additional dynamics and mechanisms. It has, along the way, fostered a metadiscourse on temporal change, an appreciation of *historicity*, not merely history, sharper in nuance and precision than ever before.

This is reflected in new metaissues that have come to the fore, such as the differing capacities of different life-forms to evolve — *evolvability*, as it is often called — and even the evolution of those differences — the evolution of evolvability (Wagner and Altenberg 1996; Wagner and Draghi 2010).

The abstract machines of selection, niche construction, and hypermediated information offer generalizations of processes central to evolutionary change. As processual patterns into which dynamics of matter and energy fall under certain conditions, they are characterized by their virtuality. Given a set of organisms with varied inheritance and limited affordances, selection is set in motion. Given the impact of organisms on their environments and the role of environments in selection, feedback interaction between constructed niches and organisms will spring into operation, altering selective possibilities. Given the constraints on the reliability of information transmission, material infrastructures for it will form and sprout rhizomatic networks connected to other infrastructures. Because of the virtuality of the machines, terms such as fitness, adaptation, and selective pressure will remain elusive.

These abstract machines go far to describe processes basic to all life. And in most life-forms, in fact, they are in principle adequate to circumscribe *all* the informational complexities involved and the whole story of their historical emergence. Something more is needed, however, to understand the further emergence of meaning in some life-forms: another abstract machine, this one conducing in its processes to sign-making — a semiotic machine. Sections 11–14 will describe this increment added to the histories of certain life-forms.

This fourth abstract machine, we will see, delimits the extent of meaning in the biosphere. Before we get to it, however, we need to pause over two alternative views of meaning, the two most developed positions in the semantic universalist camp I described in Part I. These are named *teleosemantics* and *teleodynamics*. The first has been the foremost strain in naturalist philosophies of meaning for about thirty years, while the second is a more recent, bracingly expansive vision attempting to root meaning in the fundamental nature of living things as open thermodynamic systems. These are positions, in other words, not to be taken lightly. I argue that they make wrong turns in contemplating meaning, and that these result from the confounding nature of the evolutionary *mise en abyme*.

NINE

Teleosemantics:

Trait, Function, and Meaning

Information has meaning *for an interpreter* when it is used *to achieve an end....*
An interpretation is intended for use, but an uninterpreted change simply
occurs....

Meaning is extracted from a DNA sequence, represented in the output
of an automatic sequencer, when a technician reads T[hymine] rather than
A[denine] and infers that a fetus will express hemoglobin S. The technician's
end is clinical diagnosis. Meaning is extracted from the same DNA sequence,
represented as an RNA message, when a ribosome incorporates valine
rather than glutamate into a β-globin chain. The ribosome's end is protein
synthesis.... The use of something as an object (throwing a stone), rather
than as a representation (reading a stone tablet), does not count as use of
information....

A thing has *meaning* for an interpreter when its "difference from something else" is used by the interpreter to achieve an end. An *interpretation* is a
representation of the information used by the interpreter....

Life is made meaningful by a multitude of mindless interpreters reinter-
preting the molecular metaphors of other mindless interpreters.

Questions crowd in on reading passages such as these, selected from
a recent account of meaning by biologist David Haig (2020, 243–45).
Granting to a scientist a perceived aim or telos in interpreting data
for a diagnosis, what is the telos of a ribosome operating to facilitate

the translation of a messenger RNA sequence into a polypeptide? Are all chemical and physical reactions — say, hydrogen and oxygen atoms binding to create H_2O — teleological? If not, what makes a ribosome's operation anything more than a set of "uninterpreted" changes that "simply" occur? And what about interpretation? Once we deem interpretive a snippet of RNA and a few dozen protein molecules — the components of a ribosome — have we effectively forfeited all leverage from which to explain the difference in activity between the scientist and the ribosome? What justifies the conflation of human and ribosomal meaning-making? Can we take seriously the notion of a mindless interpreter, or do we need a category of complex processes independent of interpretation? Finally, what is the nature of the link proposed here between meaning and use? Why should the aboutness of meaning be so closely linked to the achievement of an end?

Haig's account of meaning sits squarely in the philosophical tradition called teleosemantics (for an overview, see Neander 2020), and most views in this area provoke similar questions. To gauge the scope of meaning in a different way, we will want to disentangle some of the terms and issues here, distinguishing meaning from information, weighing the validity of viewing biological operations as end-oriented functions, and exploring the implications of the extension of meaning and interpretation to microbes and molecules.

I distinguished meaning from information in Part I, and I will push the distinction farther in this section and the next. To summarize my position so far: Signs are the fundamental units of aboutness, and they arise in a process that entails both a relation between sign and object and a perceiver of this relation. The resulting metarelation, binding the perceiver to its perceived world, is the interpretant. Both the metarelational construct and the perceiving entity behind it are required for signs to exist. Meaning in turn depends on signs, through which it points a perceiver toward some content about the world — past, present, or future, internal or external, real or unreal. The pointing is characteristic of the reference or intentionality basic to meaning. Information is a far broader phenomenon that subsumes meaning. All information is causal, but only a small fraction of it carries semantic content, via signs, and generates meaning. Non-semantic information is a measure of causal efficacy in the world, devoid of aboutness or meaning but covarying states at one place or time with those at another. In life-forms in their relations with their surroundings, this correlation often takes the form of signals triggering preprogrammed responses of greater or lesser complexity. Many philosophers have considered such signals to be "natural signs" (see especially Dretske 1988; Millikan 2004), but this confuses them with full-fledged semiosis and its interpretant-generating mode of mediation, not required for signaling to happen.

We might expect the idea of contentless information to come easy seventy-five years after Shannon's mathematical theory, in which there was no role for a message that might be conveyed. Easy or not, our need for the idea is pressing. Shannon's theory was part of the information turn in mid-twentieth-century thought, and biology, with its fundamental discoveries about the codes incorporated in all life-forms, was integral to this turn. Thinking about information in living things is much more than a metaphorical application to them of concepts borrowed from electronics, communication, and computation. It forms the basic epistemic framework in which we have conceptualized and organized much of what we have learned about life processes since Darwin and Mendel.

Today our formidable understanding of the transmission of information involving DNA, RNA, and mediations between genotype and phenotype in even the simplest organism highlights the division in our conception of information. All such molecular mechanisms embody causal information, but to assign them semantic content effaces an essential difference between one kind of informational mediation and a different one, whereby some few sorts of organisms create signs. An amino acid bonding with an RNA codon, or a DNA double helix unraveling partially to initiate RNA transcription, or the basic bonding complementarity of the nucleotide components adenine and thymine or cytosine and guanine: in all these there is no aboutness or reference comparable to the aboutness perceived by some animals. In relating or equating these things we ignore what is specific about meaning within the larger sphere of causal, informational biological process.

Teleosemanticists are led in this direction especially by the role in their thinking of biological functions. These are taken to be the dedicated operations of selected traits in organisms. If functions are selected, and if functions are bound somehow to meaning, then a link appears between natural selection and meaning — a pathway, ultimately, to a naturalistic philosophy of mind. Citing as their examples an array of functional traits ranging from magnetosomes in some anaerobic bacteria to human thought, teleosemanticists aim to expand questions of meaning far beyond the conditions of the human mind, beyond mind/matter dualisms, and beyond the anthropocentric questions of language, proposition, and symbolism that tend to engross philosophers of mind in the computational and connectionist traditions. This is a laudable counterstrategy in the face of humanist parochialism. The weak link, however, is the idea of meaningful functions. How is it supposed to work?

For teleosemanticists, traits are the outcome of natural selection, and they are functional because they are *selected for* their functions. Here is a typical formulation by one of the leaders in the field, Karen Neander: "It is a/the proper function of an item (X) of an organism (O) to do that which items of X's type did to contribute to the inclusive fitness of O's ancestors, and which caused the genotype, of which X is the phenotypic expression, to be selected by natural selection." (Neander 2020, §2; see also Neander 1991). Bundled together in this implicit "if and only if" formulation are many concepts basic to evolutionary studies: natural selection, function, and trait ("item") to start with, but also organism, fitness, genotype, and its expression in a phenotype. None of these terms, we saw in Section 4, come without problems of delimitation and definition. The directedness of traits that is asserted here toward particular, advantageous functions in the lives and behavior of organisms constitutes their teleology and explains the teleo in teleosemantics.

Where does meaning arise in this teleology? It is, for teleosemanticists, an automatic consequence of trait functionalism — of the purpose traits are selected for. Selected traits are systems that function in interaction with environmental signals to produce or process content, the semantics of teleosemantics. The content is conceived very widely and in many different ways, but always as some kind of correlation of environmental signals and the organismal systems necessary for survival and thriving. Every such correlation takes the form of an organismal "representation" of some aspect of the world, created by a trait designed in natural selection to produce it. The systems producing this representation, semantics, or content are not restricted to mindlike ones but include even organelles in bacteria and nucleotide sequences. Teleosemanticists explode the concept of "mental representations" that forms an ill-defined but recurrent theme for humanist philosophers of mind — another laudable goal; but they do so only by rendering representations effectively coterminous with informational covariance.

The view that traits are selected for representational functions guides in different directions two foremost teleosemanticists, Ruth Garrett Millikan and Daniel C. Dennett, and their paths can exemplify the project as a whole. Millikan begins from the "natural signs" mentioned earlier, which in her view reliably "indicate" their conditions or causal antecedents. (Already difficulties arise: Where is the perceiver of these indications? Or, how do indications exist without a perceiver?) These indications generate representations in organisms, that is, "intentional signs" created by traits selected for this purpose, and Millikan's chief concern is to understand the traits and the nature of their content, from which emerge many "varieties of meaning" (Millikan 2004). She divides representations into two types: those that describe the world to organisms and those that prescribe or direct behavior in response to the world. These act in tandem within or between organisms (Millikan 1989, 1996). Sometimes her discussions of this dynamic veer close to the Peircean interpretant, though explicit reliance on Peirce is not a significant part of her project (Millikan 1989; see Tomlinson 2018).

In this tandem scheme, the evolved traits of organisms are selected for either "producing" or "consuming" the representations. Millikan discovers intentional representations across the whole range of organisms' responses to natural signs (i.e., signals), and so she extends the domain of such representations very far; even bacteria have tandem systems, interacting with their environments as producers and consumers of representations (2004, p. 82; see also Millikan 1989). In recent work Millikan has offered a new vocabulary to describe the mechanisms of representation, one that avoids confusion of them with humanlike concepts, and she has meditated at length on evolved modes of animal function (2017). She has also had important things to say about the clumping of the world into kinds, an issue central to several strains of philosophy of biology (see Bueno, Chen, and Fagan 2018). Through all this, however, two features remain basic to her thought: the link of evolved function to content-full, semantic representations and the resulting tendency toward semantic universalism.

Dennett's emphasis on traits selected for their functions elicits from him a defense of adaptationism that has remained steadfast across several decades in the face of critiques by Gould and Lewontin and others (Dennett 1983, 1995, 2017b). Adaptations, in his view, are traits selected to fulfill the advantages they offer organisms that bear them. They are "chosen" by natural selection for their functions (1983, p. 351; the quotation marks are Dennett's). They take the form of selected, incremental design changes in organisms brought about, in one of his favored metaphors, by "cranes" doing their work by standing on foundations laid by earlier cranes (1995, chap. 3; "skyhooks," mechanisms lifting from on high and designing without the support of earlier design, are rejected; Dennett's conception is not to be confused with creationists' "intelligent design"). Evolution is the accumulation of these design changes or adaptations, and it is the task of the biologist to reverse-engineer the crane-work, thus understanding the functions that traits were selected for.

The place of meaning in Dennett's evolutionary work is ambivalent, perhaps because he wavers between epistemological positions that can be very broadly called idealist and realist. On the idealist side, a major effort in his work before he turned to evolution was to define our explanatory or interpretive stances toward the world (1987, 2017b). In this account, the "design stance," in which we attempt to understand the systematic organization of phenomena, is the stance taken by biologists in their attempt to reverse-engineer the outcomes of natural selection. The "intentional stance" instead is the one we adopt toward other minds or mindlike things: "It works by treating the thing as a rational agent, attributing 'beliefs' and 'desires' and 'rationality' to the thing, and predicting that it will act rationally" (2017b, p. 37). There is also a third, "physical" stance, interpreting phenomena according only to the laws of physics, but this leads to an unhelpful, "greedy" reductionism if applied to life-forms or mental content. All three stances are idealist in that, as one account of Dennett puts it, "there is no mind-independent determinate fact of the matter about meanings or functions," both being "dependent on interpretation, on our adopting either the design stance or the intentional stance toward them" (Neander 2020, §2). All this has been helpful in clarifying different kinds of human explanation in approaching the world.

In Dennett's many statements on meaning and its evolution, however, this idealism gains little traction. We seem to be in the presence instead of cut-and-dry, objectivist realism — indeed, it is hard to interpret his arguments with evolutionary biologists except in this way. Here meaning runs the gamut from a narrowly limited phenomenon to a pervasive fact of molecular interaction. Thus he can write: "Real meaning, the sort of meaning our words and ideas have, is . . . an emergent product of originally meaningless processes — the algorithmic processes that have created the entire biosphere" (1995, p. 427). And we can agree with him, emphasizing his proviso that the algorithmic processes are meaningless — sheer causal information. On the other hand he can speak, like Millikan, as a semantic universalist, locating meaning in every adaptation, every selected trait:

> Start with the simplest imaginable case, like a bacterium that responds to a gradient in its environment, and that response has a meaning . . . because the response in one way or another is relevant to the wellbeing of that bacterium. If it's responding to food by moving towards it, that's its meaning. . . . We have to get away from the idea that that's a merely figurative or metaphorical case of meaning. It's as real as meaning ever gets. (Dennett 2017a)

Haig is a close follower of Dennett in this regard and, as in the quotations from Haig that began this section, Dennett can carry meaning right down to the nucleotide sequences of DNA: "Through the . . . molecular-level microscope we see the birth of meaning, in the acquisition of 'semantics' by the nucleotide sequences, which at first are mere syntactic objects" (Dennett 1995, p. 204). In such statements — there are many more in Dennett's voluminous writings — the helpful heuristics and epistemic idealism of his "stances" analysis collapse into realist ontology. The assignment of meaning to selected traits is explicit, even to the most universal and basic of information-transmitting mechanisms, the nucleotide sequence and its transcription.

In both Millikan and Dennett, if in different ways, meaning is widened radically by identifying it with traits selected for the successful guidance of their organisms through their lifeways. But the automatic correlation of meaning and function at the heart of their endeavors relies on underspecified words, tokens of implicit premises, such as "representation" and "content," in addition to "meaning" itself. What are the mechanisms that re-present one thing in another, thus creating a representation? How is the deferral or displacement basic to this idea created in, for example, bacterial mechanisms? How is content, dependent on such displacement, produced and perceived? Where does it reside?

In addition to this strange, presumptive irruption of meaning, there is also a broader evolutionary puzzle thrown up by teleosemanticist positions: the question of *selection-for*. We have seen that natural selection is an abstract machine, an immanent design that gives rise to vectors of process in assemblages of matter and energy. There is no palpable, material mechanism driving the machine — no machinic infrastructure of natural selection itself, no designing cranes, only the matter and energy that fall into relations according to its schemata. Neither, of course, is there any agent doing any choosing. Natural selection entails no selecting at all, only certain conditions that set in motion, when they pertain, processes resulting in differential rates of reproduction in populations of organisms. The conditions *have* pertained ever since the earthly biosphere got started — arguably they must pertain in any biosphere anywhere — and evolution is the intergenerational shift in the overall natures of populations that has resulted.

This abstractness poses the challenge to grasping natural selection that led Darwin to his metaphorical name for it in the first place, by analogy to the choices made by human breeders in their artificial selection. It is also what made his analysis of the dynamic of selection so extraordinary a revelation. In Section 4 I likened it to Marx's contemporary analysis of the abstract form of capital, and we could extend these comparisons to include other noumenal concepts of the long nineteenth century: Nietzsche's slightly later discerning of abstract genealogical patterns where others saw positivist histories or, still later, Freud's description of the workings of the unconscious. (It is well known that all three of those thinkers were drawn into consideration of Darwin's abstraction.) Darwin's metaphor of "selection" is acceptable because abstract machines are hard to describe in literal terms. Marx spoke in a similar way of the commodity as a fetish, a metaphor aiming to capture the concealment of forces of labor and sociality at work in it. Today we speak of *attractors* in dynamic systems when nothing is attracting anything or of *emergence* in many circumstances where mediating complexities and feedback networks make it difficult or impossible to follow causal chains.

Such metaphors are useful insofar as we resist their collapse into literalness and avoid thinking of them as true descriptions — insofar, that is, as we *do not* "get away from the idea" that they are "merely figurative or metaphorical." In the case of natural selection literalness can bring especially confounding consequences that most teleosemanticists would certainly disavow. These are evident in Jerry Fodor and Massimo Piattelli-Palmarini's book *What Darwin Got Wrong* (2010), which uses a literal-minded idea of selection-for to launch an onslaught on natural selection. The authors' particular concern is the question of "free riders" — correlated features that come along with traits that are selected. If such a thing can happen, they reason, then there can be no determinate way of knowing what trait was selected for what function; hence selection-for collapses; hence natural selection altogether is an incoherent idea.

But this pseudo-proof of the incoherence of Darwin's thought rests on a literal-minded selection-for that was never a concomitant of natural selection in the first place (see Block and Kitcher 2010). In Section 4, Terrence Deacon helped us see the alternative to this common confusion regarding adaptations. Selection is an interaction of mechanisms with affordance/constraints, but no mechanism comprises only the properties involved in the interaction at any given moment. The physical features of traits are always in this way multiplex in relation to environments. The affordance/constraints, meanwhile, are never static, so at any time the multiplex nature of any trait can throw up properties newly relevant to new affordance/constraints. Of adapted functions, Deacon concludes, "few if any biological structures can be said to have only one distinguishing function. Their fittedness . . . is irreducibly systemic, because adaptations are the remainders of a larger cohort of variants selected with respect to one another and their environmental context" (2012b, p. 413). Fodor and Piattelli-Palmarini's complaint arises from a misconceived positing of determinate, isolable, nondynamic traits that allow the pinpointing of their functions. It has little to tell us concerning the role of Darwin's dynamic machine in the unfolding of earthly life.

While most teleosemanticists would reject Fodor and Piattelli-Palmarini's position, a literal interpretation of the selection metaphor and a concretizing of the abstract machine of natural selection guide their own functionalism. Again and again they assert, in a commonsensical way, that this or that trait was selected for this or that function or purpose. "Spiderwebs and beaver dams," Millikan writes, "result from the operation of inner mechanisms in the spider or beaver.... These producing mechanisms were designed pretty directly by natural selection. The purposes of these artifacts are derived from the purposes of the genes that were selected for producing them" (2004, p. 13). Dennett's functionalism depends on reverse engineering. As he says, his own definition of biology "is methodologically committed to optimality considerations. 'What is — or was — this feature *good for?*' is always on the tip of the tongue; without it, reverse engineering dissolves into bafflement" (2017b, p. 80). In this commitment to what appear to be traits, what they appear to be good for, what appear to be determinations of them by genes, and what therefore can appear to be the directional work of natural selection, we feel a slippage toward an evolutionary causality so oversimplified as to be fictional. Whatever Gould and Lewontin's excesses in their attack on Panglossian explanations, they were right to name this as the characteristic defect of adaptationism (1979). There is little sign within it of the thinking of biologists and philosophers of biology that has destabilized terms like *adaptation* and *trait* in recent decades.

Relinquishing the overly determinate notion of a trait and the idea of selection-for doesn't call into question the marvelous attunement of organisms to their environments. Just the opposite: The fluid conception of traits and functions captures the dynamic by which organism and environment constantly reshape one another, forming the ever-moving complex described in the abstract machines of natural

selection, niche construction, and hypermediated information processing. On a generation-to-generation level, the accumulation of
changes in the overall makeup of a population of organisms — the
hallmark of evolution — is not a weighing of the effective for-ness of
a particular trait in fulfilling its function or purpose but instead an
expression of this mutual flux of organisms, populations, and niches.
The question of how or at what level we might take a reliable snapshot of a trait or a function within this teeming processualism looms
large, as does the related question of how the "overall makeup" of the
group can be gauged: by genotype? phenotype? some combination?
in relation to the environment? None of these are stable or readily
tractable concepts.

These issues preoccupy biologists when they attempt to understand for-ness or function in relation to natural selection: functions
of genes, traits, social behaviors, Millikan's "artifacts," and so on (for
overviews see Perlman 2010; Cummins and Roth 2010). The problem here is not our inability to name a specific function, as Fodor
and Piattelli-Palmarini see it. Instead, biologists confront difficulties that are the converse of this as they ponder the immense range
of mechanisms, extending from molecular to ecosystemic levels,
that materialize the operation of Darwin's machine. One overview
describes succinctly the difficulties this entails: "It may be that the
biological world is complex enough that we will be unable to find
necessary and sufficient conditions for some biological character
trait to have a function . . . or, for that matter, for something even to
be a biological character trait!" (Pigliucci and Kaplan 2006, p. 134).

This destabilizing of traits and their functions has elicited from
evolutionists many alternative approaches. Sometimes these emphasize the epistemic provisionalism of our conceptions, for example, the
"modern history" view in which only the most recent maintenance
of a trait, not long-term evolutionary directions, can be judged useful
in assigning function to it (Godfrey-Smith 1994; Griffiths 1993). The
presentism of this approach tacitly admits that historical outcomes of
evolutionary processes cannot be captured or explained in functional

snapshots. Other approaches are more realist in orientation, focusing less on epistemic concerns. In the "developmental systems" or evo-devo view of organisms, for example, genes and traits are provisional entities in dynamic systems, plastic in their operation across the lives of organisms (Oyama 2000). In a new view of homology arising from evo-devo thinking, homologies are not ancestral traits varied according to selective demands for differing functions but instead differing expressions in differing environments of the same, long-conserved regulatory networks (Wagner 2014). In the niche-constructive view, traits are fluid epiphenomena of the organism/niche matrix (Odling-Smee, Laland, and Feldman 2003; Lewontin 2000). In the broadest view of all, finally, function is synonymous with the feedback regulatory networks fundamental to all living things (Keller 2010). It is the operation, in other words, of the third abstract machine of evolution, hypermediation. All these initiatives approach the dynamic of evolutionary change and stasis without relying on overly reified, finally fictive adaptations and functions.

As function and traits grow more mutable and difficult to define, the plain for-ness of selection evanesces, and with it the indications of traits-for, their representational capacities proposed by teleosemanticists. This severs the structural relation they posit between meaning and functions as products of evolution; the complexities biologists wrestle with are seen to fall into the category of Dennett's "meaningless processes." Meaning resides elsewhere, in short. It can supervene on evolution only through additional complexity arising from the nonteleological operation of selection, radical niche construction, and hypermediated information processing — specifically, from a folding over onto itself of the interaction of organism and niche that structures information as a particular kind of relation to a relation.

We have circled back to Dennett's meaningless processes — as we must, since his bottom line and ours is natural selection — but with a new sense of the transformations possible in the crane-work. Meaning is neither ubiquitous and coextensive with information, nor is it what is indicated by selected function. (Nor is it limited to human

minds, the parochialist position.) Instead, it is a novel kind of outgrowth from meaningless processes in a small area of the biosphere.
And it is a puzzling outgrowth at that, since most life-forms reveal
that sheer, meaningless information can frame niche/organism interactions of immense complexity, success, and durability. How did the
meaningless processes give rise to meaning? What are the biological
mechanisms that kick-start these processes? How large is meaning's
range, and with what creatures do we share it? Before setting out to
answer these questions, we must take on the most comprehensive,
venturesome, and seductive extension of meaning throughout the
biosphere offered in recent years: Deacon's teleodynamics.

TEN

Teleodynamics:

Constraint, Work, and Meaning

Teleodynamics is the name Deacon gives to a particular kind of emergent system of systems, an elaboration of thermodynamic principles enabling both self-structuring and maintenance of the self-structuring capacities (Deacon 2012b). These are fundamental features of all life-forms, so Deacon's conceptualization amounts to an ambitious thermodynamic description of life, placing him in a line of thinkers reaching back through the systems theorist Ludwig von Bertalanffy (1969) to Erwin Schrödinger in the 1940s (2012). In its explanation of life from the vantage of fundamental physical laws, Deacon's position might seem a "greedy" reductionism in Dennett's sense. If greedy, however, it is not reductionist, because it builds also on postulates of complexity theory and dynamic systems theory advanced by theorists of life such as Stuart Kauffman (1993) and Humberto Maturana and Francisco Varela (1980) in order to create a nonreducible model of an emergent phenomenon. In a labyrinthine, brilliant presentation, Deacon carries this thinking farther than anyone before him, tracking the appearance of life through a hierarchy of systems of increasing thermodynamic complexity.

But Deacon's ambitions don't stop at life alone. His ultimate goal is to explain significance, sentience, consciousness, and mind as phenomena likewise emergent from his systems dynamics and teleodynamics. He is a natural philosopher of meaning of a kind more intent upon fundamental, physical principles than the teleosemanticists, and his semantic universalism is for this reason still more sweeping than theirs. In relation to meaning, however, his ambition proves to be more than his model can support. Isolating some of the steps of his argument can show us how this is so.

Deacon's model builds from three fundamental elements: the tendency of all isolated systems to move toward maximum entropy (the second law of thermodynamics); the energy-consuming work done by some non-isolated systems, which produces change in the direction of decreasing entropy, that is, increasing order; and constraints, extrinsic or intrinsic restrictions of systems within certain limits or states, which impose conditions for their work. Deacon employs these factors to describe a nested hierarchy of levels of thermodynamic complexity, starting from a level of systems described by the second law in which any change moves spontaneously from order to disorder, from lesser to greater entropy. Two higher levels emerge from this one, manifesting new kinds of relations of constraints and work.

At the next level, *morphodynamic* systems are self-organizing, generating increasing order through work under conditions of constraint (2012b, chap. 8). Their self-structuring maintains them far from thermodynamic equilibrium with their surroundings. The energy to maintain this non-equilibration comes from outside the system in the form of external perturbations, and the dissipation of this energy is what enables the building of order in the face of the second law. The resulting structure is highly dependent not only on the structure-building mechanisms of the system but also on the nature of acting constraints, which thus prove to be negative complements of order. One of Deacon's examples for morphodynamic ordering is the formation of a snowflake, in which energy for the structuring comes from the fusing of new molecules into the crystal, which releases heat, and each stage of crystalline structure constrains the possibilities for further structure. The order of the flake is self-generated under the conditions of these energetic accretions, involving extrinsic constraints, and according to the additional, intrinsic constraints on crystal growth (pp. 257–59). A kinship is evident between this example and those of streambed sedimentation and granite formation, used in Section 5 to exemplify de Landa's abstract machines. Deacon's morphodynamics can be thought of as an elaboration of de Landa's attractor, sorting, and meshwork machines, all of which exploit the input of energy to generate order.

Teleodynamic systems form the third and final step in Deacon's thermodynamic hierarchy, after entropic, second-law systems and morphodynamic ones, and they cross a phase transition into a new kind of organization (chap. 9). This marshals morphodynamic self-structuring systems into systems of systems of which the output is not only structure but *the structuring of structures that enable structure.* The tautology of wording is almost inevitable in the face of the recursive shape of teleodynamics, and this recursion makes the systems not only self-organizing but also self-maintaining. It emerges from a compounding or coupling of morphodynamic processes in which the work/constraint relation of one is pitted against that of another, forming a higher level of working possibilities. At this level unanticipated attractors form that become endpoints for teleodynamic processes; tending toward the endpoints, the processes create self-maintaining and self-generating capacities. This tendency offers an initial explanation of the teleo of Deacon's term — though he will put it to further work, as we will see.

With his teleodynamic systems Deacon approaches biological reproduction and lineages of systems. Mere self-maintenance of morphodynamic components is not sufficient for their persistence, and there is an "incessant need to replace and reconstruct organism components" that "depends upon [morphodynamic] form-generating processes" (p. 276). The tipping point into life and its varied reproduction is at hand — the point at which natural selection springs into operation and de Landa's "blind probe head" machine appears. Deacon's touchstones in describing this point are the autopoiesis of Maturana and Varela (1980) and Kauffman's autocatalysis (1993), and he elaborates on these in describing the "autogenesis" of not-quite-alive, "minimal teleodynamical systems" resembling the protocells thought to have played a role in the origins of earthly life. These autogens (as he names them) are bounded, self-assembling and self-perpetuating entities consisting of lipid membranes or protein shells enclosing concentrations of molecules likely to enter into mutually catalytic, self-perpetuating interactive cycles (chap. 10).

Here a first, audacious step occurs in Deacon's argument—a move that will ultimately carry us (via several more) to his view of interpretation, reference or aboutness, and signs and meaning. Because teleodynamic systems of systems bring about new attractors, Deacon judges, they are end-directed: "Teleodynamic processes can be identified with respect to the specific end-directed attractor dynamics they develop toward. . . . Teleodynamics is the dynamical realization of final causality, in which a given dynamical organization exists because of the consequences of its continuance, and therefore can be described as being self-generating" (p. 275). The putative causality here is a complex one, starting from Aristotle's final cause, "that 'for the sake of which' something is done" (p. 34). For Aristotle this implied a human agent—a carpenter building a house, in Deacon's example, with a prospective vision of the result of the work. But there is a retrospective element also to the Aristotelian analytic of different causes, a view from which an interpreter of a situation can look over its unfolding and distinguish final from material, efficient, and formal causes. In either case, prospective or retrospective, a purpose, goal, endpoint, or telos is identified toward which the work points.

In the light of the discussion of teleosemantics, we can see here a difficulty taking shape. Assigning a telos to any teleodynamic system means that, in the case of life-forms—all teleodynamic—we are cast back into the realm of purposive biological functions. Asserting that teleodynamic systems in living things exist "because of the consequences of [their] continuance," Deacon in effect hypostasizes the traits-for and functions of Dennett, Millikan, and others into a universal thermodynamic model of life: a *thermodynamics-for*. The problem of for-ness in living things does not disappear in this move but is instead concealed in the all-important phrase "because of." How does the fact of an attractor in a system rebound on that system to render its existence because-of? How is such a system like a carpenter with a plan?

What is needed, Deacon argues, is an interpreter not restricted to the human — a dramatic widening beyond Aristotelian humanism. Deacon seeks it in the putative for-ness of teleodynamic systems, but this requires a second step. The end-directedness of teleodynamic systems describes for Deacon the defining condition of interpretation. "We can describe interpretation," he writes,

> as the incorporation of some extrinsically available constraint to help organize work to produce other constraints that in turn help to organize additional work which promotes the maintenance of this reciprocal linkage.... The interpretive capacity is thus a capacity to generate a specific form of work in response to particular forms of system-extrinsic constraints in such a way that this generates intrinsic constraints that are likely to maintain or improve this capacity. (pp. 398–99)

Replace the word "interpretive" in the second sentence with "teleodynamic" and you get a definition of teleodynamics itself. Interpretation is teleodynamic, teleodynamics interpretive. We can see here Deacon's attempt to define interpretation thermodynamically — a move that goes beyond even Haig's interpretive ribosomes and molecules. Evolution "generates the capacity to interpret something as information" in a way that is "intrinsic to a self-perpetuating, far-from-equilibrium system" (p. 416). This is a large step — a true leap, in fact. Across it, interpretation emerges as an automatic outcome of any self-regenerating work in response to an environment, and so *all life-forms interpret*. Trees in their seasonal changes interpret "likely future events" through adapted mechanisms that have acquired "interpretive reliability" (p. 417), lineages of autogens are not merely selected but "selective about which environments are best to dissociate and reproduce in" (pp. 442–43), and genetic information "is interpreted by the persistence of the self-perpetuating process that it contributes to" (p. 416). But the first example assigns prospective causality where there can be none; the second seems to intimate an actual choosing in natural selection — the very confusion Darwin saw in his term but knew to avoid; while the third renders all genetic transcription interpretive, not to mention mysteriously installing the interpretive function in such processes themselves. A semantic universalism worthy of John Maynard Smith (see Part I) is taking shape.

A third step brings us to representation. This, Deacon avers, repeating a mainstay of poststructuralist thought, is founded always on a constitutive absence since it involves the presenting over again of something else in some aspect. For Deacon, this absence is congruent with the extrinsic constraints upon which both teleodynamics and interpretation depend. Constraints too are not present in the systems that work with them to create order. And if that is so, then representation or aboutness is another manifestation of the coupled constraints basic to teleodynamic systems. "A fit or interdependent correspondence between constraints in different domains," he concludes, "is the essence of both biological adaptation and the relationship characterizing representational relationships" (p. 324). Function and representation are congruent thermodynamic processes, "made possible by the way living processes are intrinsically organized around absent and extrinsic factors" (p. 418). Just as teleodynamics itself is interpretive, so biological function is representational. Once again Deacon's argument veers close to the teleosemanticist position, with its representations discovered even in bacterial magnetosomes and its pervasive functional aboutness of traits-for.

A fourth step, finally, introduces signs, the hallmarks and foundation of aboutness. We saw in Part I that Deacon is a practiced Peircean semiotician, and his writings on Peirce date back to the 1990s, well before his teleodynamic theory took shape (see 1997). Here that expertise comes back into play, as he tries to derive semiotic processes directly from his thermodynamics. Signs, Deacon knows, arise through complex relations that are not reducible to any one of their three component parts: object, sign vehicle, and interpretant. Signs are, like the representational relation they instantiate, constituted from absences through the formalization of relations (and, as we saw with Kockelman, metarelations). For Deacon, this draws signs — again, it seems, by dint of sheer formal congruency — into his general schema concerning work and the absent presence of constraints that enable it to create, under teleodynamic conditions, self-generating and self-perpetuating structure. So as he sees it, *signs are automatic concomitants of teleodynamic systems*, like interpretation and representation before them. Their aboutness and meaning extend across all of life.

Semiotics reaches even to Deacon's hypothetical minimal teleodynamic systems, his choosy autogens (2012b, pp. 442-46). These can be supposed to be "sensitive" to the substrates in their environments, binding with them and breaking their containment membranes when the substrate ingredients will increase the concentration of autocatalytic molecules inside them, otherwise resisting such breaching. In this way, an autogen can be said to "respond selectively to information about its environment." (Remember: it is not natural selection acting on autogens that is mooted here, but a molecular mechanism of selection *by* autogens.) Aboutness makes a quiet entrance here, where simpler, morphodynamic covariant states of molecular interactions would seem to be sufficient; but it will not stay quiet for long. The autogen is now, in Deacon's view, actively interpreting its environment. It must be so since it is a teleodynamic system, and interpretation and teleodynamics have (in the second step described earlier) been rendered coextensive. The autogen's binding of certain molecules is directed toward the telos of perpetuating its autocatalytic self-generation.

In this interpretation Deacon discerns all three elements of the sign. The sign vehicle is the binding of substrate molecules to the autogen membrane or shell, the object is the "suitability of the environment" somehow represented by the binding, and the interpretant is the consequence of the binding — "the maintenance of . . . interpretive capacity" or the perpetuation of the autogen's "interpretive habit" (p. 443). This, then, is Deacon's final move: even not-quite-alive autogens *automatically assemble a whole Peircean sign system*. And, if autogenic processes are inevitably semiotic in this way, so are all processes of fully living organisms. Deacon here carries signs and their meanings as far as Dennett or Haig, with their meaningful nucleotide sequences, and like them he eradicates the possibility of contentless information in the biosphere. The general import of this extension with respect to natural selection is dramatic: all evolved operations, in principle all molecular interactions in every living organism, are teleological, semiotic, and meaningful.

Deacon worries how there can be such a difference between his molecular autogen, fully semiotic, and a nonsemiotic mechanical sensor like a thermostat, and the worry is well placed. He sees the difference in the "necessary intrinsic normative character" (p. 446) of the autogen's interpretation and proclaims normativity to stand behind all interpretation, distinguishing it from "a mere causal process." It entails serving some purpose or promoting a "favored consequence . . . that is in some way valued over others" and is possible only in teleodynamic systems, with their direction toward an end. To interpret is to support the normativity of a teleodynamic process, and "to explain the basis of an interpretation process is to trace the way that teleodynamic work transforms mere physical work into semiotic relationships, and back again" (p. 393). The introduction of the ideas of normativity and value seems arbitrary here, when autogen reproduction is a question of statistical tendencies of certain molecules binding with certain other ones in a molecular soup; it is as if some humanistic special pleading is called for to narrow the gap between autocatalysis and the sign. My view of the distinction between causal and semantic information answers the worry in a different way: Both thermostat and autogen embody only causal processes, but there is, for the hypothetical autogen, no "mere" about the causality. It manifests the beginnings of the hypermediation and niche construction that will characterize all life-forms.

In sum, four moves carry Deacon to a meaning that pervades the biosphere:

- assigning teleology to teleodynamic attractors;
- extending interpretation to all of teleodynamics;
- identifying function with representation under the aegis of constraint; and
- discerning the three-part Peircean sign in all teleodynamic processes.

The overall effect of these, in keeping with Deacon's universal ambitions for his thermodynamic model, is to assimilate all processes of meaning-making to teleodynamics. And, though teleodynamics is in itself not a reductionist model, this assimilation enacts a greedy reductionism on interpretation, representation, semiosis, and meaning. Meaning is flattened and thus extended to radical lengths in the biosphere, far beyond anything we can connect to the presence of content in minds or cognitive systems.

Lost under this expansive umbrella are all the differences a mapping of the habitations of meaning in the biosphere should seek to understand: emergent differences that characterize the attunements of certain organisms to their niches and distinguish them, through sign and aboutness, from the complexities of other organisms' attunements. The same loss characterizes teleosemantics, with its meaningful functions assigned to the traits of all organisms. The conception of Peirce's sign I introduced in Part I, unlike Deacon's, can help us not to efface these differences while also not limiting meaning to a single, upright animal among all others. It can chart meaning's reach across a range far larger than humanist parochialism does, without lapsing into undifferentiated semantic universalism. It's time to place this conception in the context of the first three evolutionary machines described earlier.

ELEVEN

The Fourth Machine: Semiosis

The range of meaning is delimited by capacities and processes that enable animals to perceive relations to relations in the world, constructing thus the metarelation essential to signification and aboutness. In evolutionary history these animal abilities were not selected for sign-making. They took shape instead in niche-constructive negotiations that altered, in incremental ways and without for-ness, the web of mediations between organisms and environments, extending from molecular to ecosystemic levels. The metarelational niche construction that resulted marked the advent of a fourth evolutionary abstract machine: the semiotic machine.

Judged against the whole of the biosphere, the sphere of meaning is narrow, generated by a group of animals that, though sizeable, is dwarfed by the totality of earthly life-forms. A first question arises: If the semiotic machine is of such limited scope, and moreover derivative from the three grand machines that pervade and even define all life-forms, why does it merit association with them? Or, if semiosis counts as a basic machine, why not enumerate many more? We might consider photosynthesis an abstract machine, or energy-producing metabolic pathways conserved across huge stretches of life, like the Krebs cycle, or signal transduction circuits leading from stimuli through receptors to those metabolic pathways — all processes far more widespread than semiosis in the biosphere and, like semiosis, emergent from the first three abstract machines. But here is the difference: These processes, for all their crucial roles, did not alter the fundamental nature of those more basic machines. Semiosis, however, introduced a new kind of channeling of information. Because of this it brought about a qualitatively different kind of niche construction, and in changing that machine it altered the effects of the other two. These alterations justify ranking it among the major transitions in the evolution of earthly life (Maynard Smith and Szathmáry 1995) — a topic I will take up again in Part IV.

The new kind of information transmission enabled some animals to build new kinds of niches, crisscrossed by representational connections. A novel process was added onto the general dynamic of niche construction common to all life-forms, one that is, in a sense described in Section 13, *analytic*. Its effect is to introduce folds in animals' experiences of the world, bringing into proximity things that had existed before only as independent phenomena or disconnected stimuli. The effect ripples out to alter the affordance horizons of semiotic animals, reshaping not only their niche construction but also that of other, nonsemiotic life-forms. The fourth machine can engage in new ways all the processes involved in the radical niche construction discussed in Sections 6 and 7, at all temporal and spatial levels, from moments to eons, molecules to ecosystems.

In its renovation of niche construction, semiosis might appear to be an independent force, external to that organism/environment relation. But this is an illusory independence, since all its processes operate within and as part of the embedded mediations of the animals capable of it. *All semiosis, all signs, and hence all meanings are niche constructive*, and this carries straight through to *Homo sapiens*, since whether a sign produces an immediate enactment in the world or not is irrelevant to its niche-constructive nature. The highest flights of human fantasy or concept remain part of an organismal stance toward the world, helping to define its ongoing negotiation with its niche, and even the most complex social contexts for such flights are mediating layers between them and their niches.

TWELVE

Abstract and Episodic Signs

Let's return to Peirce's understanding of the sign (see Part I). An icon is a sign that bears a qualitative likeness to its object, an index a sign that points to or indicates its object, often through proximity to it or causal relation with it, and a symbol a sign related to its object through generalized conventions, rules, and the like. Peirce discerned these types of signs along one vector in the triple mediation of the signifying process, the one relating the sign vehicle to the object and picking out specific features of it that enter into the metarelation. Along a second vector, the reverse of the first, the object exerts its own determining force in the semiotic process, delimiting the possibilities that can be picked out and constraining the nature of the sign vehicle. And the interpretant, the calling of an organism into this system of constraints and its response according to its capacities, adds its additional determinants of signification. Constraints from all sides mark off the partial or aspectual nature of signification; this is a defining trait of the semiotic metarelation and hence of all aboutness and meaning. Beneath this process, in Peirce's formulation, lies the *ground*, which he took to be something like an idea, a mind, or deep-formed habits of mind. Here we will instead understand it, in a more biologically neutral and accessible sense, as evolved processes in certain cognitive entities — entities exploiting a neural network — that mesh to form an interpretant.

Peirce's three sign types embody three distinct ontological possibilities for signification: qualitative resemblance (icon), effective or pointing connection (index), and rule-governed specification (symbol). These types array themselves in an *a priori* hierarchy of dependency on one another, such that indexes are dependent on qualitative likeness, hence on iconism, and symbols rely on the pointing of indexes to define their law-abiding interrelations (Deacon 1997 and 2012a; Tomlinson 2018). These ontological distinctions encounter animal capacities at work in the formation of the interpretant, together determining the kinds of signs animals create.

Among the three sign types, the *icon* is the most basic but, just for this reason, hardest to ponder. Its interpretant fixes on some quality shared by the sign vehicle and object as the relation between them, but — a first complexity — this can range from a simple and unitary quality — greenness as an icon of grass — to a similarity embracing a constellation of many qualitative features (Peirce 1955, pp. 106–107). When a Paleolithic human depicted a mammoth on a cave wall, the representation might need only a twice-curved line, capturing the shape of a humped head and shoulder. Or it might require much more; to paint, for example, a leg alone might not distinguish it from a rhinoceros, and the single curve of a tusk might convey any number of other curved things. In this case, the icon involved combining in proportionate relation more than one essential quality. Like all signs, these mammoth icons were partial and aspectual. Omitted from them were other qualitative features that could have been included, such as color and wooliness, not to mention features beyond the medium of paint, such as scent or sound.

We can describe what appear to be similar icons formed by a bird, though they will be provisional and hypothetical, for reasons we will see. When a bird recognizes another bird in the distance as predator or prey and not a conspecific, it might do so because of a single qualitative aspect: size, shape, color, the motion of its wings, its call, and so forth; or it might take account of the relation of several aspects. Birds can learn foraging tactics such as the nature of foodstuffs and the sites where they can be found by observing their parents or other birds. House sparrows in captivity have been shown to learn a preference for food colored a nonnatural yellow by observing conspecifics eating yellow food (Friday and Greig-Smith 1994). Great tits and blue tits cross-fostered — reared as hatchlings by parents from the other species — learn to choose larvae to feed their own young that conform in dimension to those of the foster species, though the two species otherwise forage different-sized prey (Slagsvold and Wiebe 2011). We can identify — provisionally, again — the yellow of the sparrows and the larval size-range of the tits as icons representing what each bird eats, joining with whatever innate, non-semiotic signals are involved in their feeding and foraging.

The distinction of avian signs from signals is important, and a contrasting case from honeybees, involving a honeybee's sensitivity to sucrose concentration in the nectar it collects, can clarify it. High concentrations of sucrose induce a cascade of neurochemical and behavioral changes in a foraging bee. A single, long, sucrose-sensitive neuron begins the cascade, projecting to reward-processing neurons, sensitive to increases in this most basic foodstuff, and motor control neurons. The neurotransmitter involved in this network forms, through associative, Hebbian learning, basic memories of the food sources the bee has visited, consisting of neurons that once fired together firing in synchrony again (Giurfa 2007). The neurotransmitter also induces arousal, which leads, when the forager returns to the hive, to enhanced sharing of regurgitated food with other bees (Gil and De Marco 2005). The sharing sets off similar cascades in them, creating neural firings that bring about their own associations. These either lead them to seek the first forager's source (if they are inexperienced foragers) or stimulate them to revisit sources they have already memorized (if they are old hands). Here there is no interpretant and no sign, only the signal functioning of an attuned neurochemistry and resulting behavioral concomitants.

Of course, neurochemistry also underlies at a foundational level the learning of birds, but these two instances of learning are distinct. The bee's learning involves no observation and processing of a life situation or *episode* — a word that will be important for us — but results from direct chemical stimulation of a neural net. It might require no percept at all, only a sequence of innate responses to chemical concentrations. The learning of the sparrow or blue tit instead leads, through many levels of neural mediation, to the perception that certain elements in an experienced episode stand in relation to what counts as food. Learning of this complex kind and the ability to memorize and retrieve whole life episodes are capacities an animal must possess in order to form signs.

An *index*, according to Peirce, is

> a sign, or representation, which refers to its object not so much because of any similarity or analogy with it, nor because it is associated with general characters which that object happens to possess, as because it is in dynamical (including spatial) connection both with the individual object, on the one hand, and with the senses or memory of the person for whom it serves as a sign, on the other hand.

Indexes are different from icons and symbols in three ways, Peirce continues: by having no resemblance with their objects that counts as part of the signification (even though they may share qualities with it); by referring to individual "units" or events, or to a coherent collection or continuum of them; and by directing attention to their objects. "Anything which focuses attention," Peirce pronounces sweepingly, "is an index." The dynamic relation of index to object revolves around issues of spatial or temporal proximity, of a pointing or indicating relation, and of causal relation — an index, Peirce writes, refers to its object "by virtue of being really affected" by it (Peirce 1955, pp. 107–108, 102).

Important implications unfold from these succinct descriptions. Unlike an icon, an index depends for its signification on its surroundings or context, to which it might be bound in several ways. This contextualism is of a piece with indexes' situational nature: they are signs created in relation to single, specific instances or events, however multiplex or complex these might be (hence Peirce's qualification about collections and continua of events). Their pointing function is also connected to their contextual nature since it enables them to specify the situation in which they arise. It is this pointing, finally, that engages the attention of the animal registering the sign.

These features render indexicality an especially diverse, flexible, and versatile kind of sign-making in the lifeways of semiotic animals. They also blur the distance Peirce wished to maintain between icons and indexes, at least for nonhuman animals. This is because of the relation to a situation or episode from which an index derives its meaning. Let's consider the human situation first. A mammoth drawn on a cave wall represents iconically all mammoths, but the same proportion of head and body seen in the distance, like a scent picked up on the wind or trumpeting heard from afar, represents a specific mammoth, in a particular situation. The bodily designs seen on the wall and in the wild seem to be similar signs, but the situational context differentiates them in a way that is as crucial to our consideration of semiosis as it could well have proved to a Paleolithic tribe. The individuality of the object, its specificity in the here-and-now (or in a remembered or imagined here-and-now in past or future), alters fundamentally the sign. It engages the sign in the lived experience of the sign-making animal, pinning it to a moment-to-moment situation and to the bits and pieces of a perceived world that cohere in it.

We must imagine that the people in the cave, given full-fledged memories of earlier incidents or projective imaginings and stories of future ones, might convert the icon on the wall into an experienced index. But we also suppose that they had a capacity for abstraction that enabled them to create, in the painting on the wall, a true icon, a representation of all mammoths, any mammoth, or simply Mammoth. This is part and parcel of a swelling "release from proximity" in the hominin imagination across the last half-million years that paleoanthropologists have tracked through many archaeological proxies (Tomlinson 2015).

The evidence marshaled in Part III will suggest that a bird creating a sign has little or no capacity for such abstraction. (I'll qualify this blunt statement in several ways without reducing its importance.)

This means that the experience of pure iconism is out of the reach of birds, and that they process all icon-like signs as signs created in situations unfolding in their lives — as indexes in life episodes. This is why my descriptions of bird icons needed to be provisional, hypothetical. The movements, shape, or color of another bird might be said to represent, qualitatively and hence iconically, a predator. But the bird does not conceive Predator or Hawk in a general sense, and every semiotic experience of one or more of those features is part of a real-life encounter with a predator that, in its episodic singularity, renders the signifying features indexical. The sign becomes a pointer to a situation of threat, a signifier reliant on its context for its meaning, an attention-grabber of the first order, and a reflection of proximity and causal connection. In the same way, we will see, *we* can think of certain gestures in birdsong as something like icons of particular qualities — "sexiness" in the case of canaries, for example. But for a canary these only ever signify in real-life contexts that convert them into mobile, flexible, dynamic indexes.

The reasons for this limitation of iconism have to do with the cognitive capacities available to the bird for semiosis, which do not enable it to abstract qualitative features from episodic experience and bind them to an object in a pure, nonsituational iconism. The same seems to be true for all other nonhuman semiosis in the world today (though, as with birds, qualifications are no doubt necessary in certain cases): beyond humans, icons are rare or nonexistent, while the index is ubiquitous. This restriction to indexes does not make nonhuman semiosis ineffective. In its marking of connections in situational contexts, indexicality is eminently functional for the animals involved — all they require for their meaningful niche construction.

The icon, then, can be considered the most basic sign only in a typological way and according to its essential, basic, qualitative nature. In lived semiosis it demands for its full realization a degree of abstract cognition that only humans seem to possess. (Peirce, for his part, reflects the abstractness of icons in his choice of examples of them, which include such human conceptualizations as diagrams and algebraic formulas.) For a tiny thought experiment, imagine — though this cuts against conclusive archaeological evidence — that Paleolithic painters worked with cognitive conditions similar to those of birds and other nonhuman semioticians, unable to form abstractions. The questions that arise are: Why paint at all? And on what cognitive foundation? A conversion of the questions into hypothesis follows: Ancient humans, sapients and perhaps other groups also, painted pictures of things *because of* their abstracting powers. In their actions iconism became, for the first time in the earthly biosphere, not a metaphysical *a priori* but a real practice.

The border we are concerned with between semiotic and nonsemiotic realms is populated, close in on the semiotic side, by indexes as well as icons *in potentia*, rendered indexical. *Symbols*, signs that derive their meanings from sets of rules, laws, and conventions, are probably a uniquely human kind of semiosis in the world today; at most, they are rare to the point of nonexistence in the nonhuman world. Nevertheless, there is one path along which some nonhuman animals draw near to symbolism.

The signs in an elaborated symbol system relate to one another according to the rules binding them, which entail the grouping of signs into classes with specified modes of connection among them — the parts of speech of natural human language, in the foremost example (Deacon 2012a). The rule-governed relations of these classes form syntaxes that, in language, yield the possibility of propositional structure: noun + verb +.... Without being arranged in syntactic relation to one another, individual words convey meanings only in the abstract. The difference between *tree* and *this tree* or *the tallest tree* or *the tree that fell last night* is all-important in communication, since it connects meaningful words out to the world — phenomenal, imagined, remembered, predicted. This distinctively human difference replays again and again the Paleolithic difference between *mammoth* and *that mammoth over there.*

This specification or binding among words and between words and the world are two pointing functions involved in language, which is to say indexical operations that language as a symbol system is founded on. Indexicality thus determines the dependency of words or phrases on their syntactic and worldly contexts for their meaning. Linguists refer to such contextual dependency as *deixis*, and it's no accident that Peirce often turns to pointing words or *deictics* — pronouns, prepositions, quantifiers of several sorts, and more — when he wishes to exemplify indexes. Such words function particularly clearly in binding linguistic meaning to the world.

Nonhuman animals can also muster syntaxes in their communication, as we will see in Part III, but these do not function exactly in the manner of human linguistic syntax. They don't specify links between semantic contents of their individual components in the manner of linguistic propositions. Indeed, these components mostly seem to have no semantic content, raising important questions about the role of the syntactic organization itself. In addition, the components of nonhuman syntaxes seem not to fall into classes to which rule governance applies. There seem to be no parts of speech in birdsong or whale songs. These nonhuman syntaxes are, finally, not symbolic, but instead organize to one degree or another the indexes created by the animals building them. In examining ancient hominin communication reaching back before the time of sapient humans, and before the time of modern language and symbolism, I have termed such ordered arrays of indexes *hyperindexical* (Tomlinson 2018). The term is applicable also to the indexical syntaxes some nonhuman animals generate in the world today.

What are the cognitive capacities necessary for semiosis? Answering this question, even in a preliminary way, will require the remainder of Part II and all of Part III, where I discuss two sorts of animals that sit on different sides of the semiotic/nonsemiotic divide. The situational nature of indexicality suggests a starting point, for it depends on advanced sorts of learning and memory. The interpretant formation it involves needs to locate the object as an effective element in a remembered episode and to connect it to the sign vehicle within the experienced coherence of that episode. This entails grasping a situation as some kind of unitary thing and at the same time perceiving its components and their relations: a complexly hierarchic grasp, both synthetic in its holism and analytic in its parsing of the perceived wholes. The lessons learned from the synthesized, analyzed episode are essential to index-making. To make smoke an index of fire, in the most famous example of all, an animal must have experienced smoke and fire together in an earlier episode; it must remember the episode and its elements; it must be able to retrieve the parts in relation to the whole; and it must have learned the threat that the smoke indicated. (The matching of the remembered episode with the present situation is an iconic element within the structure of indexicality, just as words pointing to one another or to the world reveal an indexical element in symbolism — one more indication of the hierarchic relation of sign types; see Deacon 2012a.)

Behind these powers of memory and learning stands a still more basic cognitive capacity: attention. Anything that focuses the attention, Peirce proclaimed, is an index. But what *is* this focusing, and how do some animals manage it differently from others? Attention, memory, and learning of certain, advanced kinds: these are the foremost requirements for the semiotic machine to crank into operation.

THIRTEEN

Attention, Situational and Analytic

Picture an animal capable of being *attentive*. Common sense tells us there are very many such animals in the world today: crows, lions, chameleons, swordfish, octopuses, and so forth. Whole groups of animals — birds, mammals, reptiles, fish, cephalopods, and more — seem united in their attentiveness. The same common sense suggests that many other groups of animals, for example, flatworms, clams, and sponges, are at best weakly attentive, and more likely inattentive, without any capacity for attention. Insects and many other invertebrates may inhabit a middle ground. These rough-and-ready estimates suggest a spectrum of attentional capacities, with different animals showing more and less developed forms. Among vertebrates, where most research on attention has been focused, selective attention, the targeting of some things over others for attention, is widely dispersed, at least in rudimentary form (Krauzlis et al. 2018). But a spectrum of differences is also evident: birds and mammals pay visual attention in more varied, flexible, and self-controlled ways than reptiles, amphibians, and fish, while primates show further subtleties absent in other mammals (Knudsen 2018, 2020).

These differences bespeak long evolutionary histories, and in the longest *durée* it seems likely that the beginnings of attentional capacities reach back to some animals in the seas of the Cambrian period; perhaps they enabled the first targeted predation there (Haladjian and Montemayor 2014). But evolution's abstract machines have wrought changes in the nature of attention across much shorter spans. Significant differences have been detected, for example, between attention in humans and macaque monkeys, separated by only 25 million years of evolution (Patel et al. 2015). These might reflect, the researchers propose, the usefulness of different attentional capacities in distinct lifeways: quick processing on the macaque side, enabling micro-moment responses in arboreal life, and slower evaluation of subtleties of social interaction on the human side — the reading of facial gestures, for example. On the human side, the differences are generated in brain structures not found in macaques, hence presumably not in our last common ancestor. These suggest general correlations between attentional capacities and the structures of neural systems and intricacies of their connectivity, a topic we will come back to.

Attention rises on a foundation of organismal response to the world, but it is something more than a response alone. Responses to signals, internal and external, are basic to all life-forms, an aspect of the transmission of information by which they maintain and regulate themselves. Across the biosphere, most such responses, complexly determined though they may be, do not involve attention or indeed any kind of perception. The cellular dynamics that lean a flower toward sunlight, the movement of a bacterium toward higher concentrations of a nutrient, the closing of a sea anemone's tentacles around prey that moves too close, or changes in my blood circulation after a meal all arise from intricate channelings of information, but none of them entail attention. Attention involves something added to this: the ability of a neural system to direct itself to or *focus* on stimuli selected from all the information bombarding an animal at any moment. From the sun on its skin, the breeze in its fur, the rustling of a branch, the movement of prey or mate, the scent of a conspecific, the twitter or bark of a nearby animal — from all this and much more an attentive animal can winnow out one stimulus and bring it into special salience. Attention is a focused perceptual awareness. Most biological signaling is not.

This focusing need not begin as an act of conscious choice;

probably it mostly does not. We know from our own experience that our attention can be captured without our willing it to be so. Working from a store of long-term memory, I can consciously direct my attention to a task at hand, as when I turn to my work at the beginning of a day. But the blackbird singing outside my window captures my attention in a different way, and the idiom of *attention captured* conveys well the involuntary outset of the experience. This dichotomy of involuntary (*exogenous*) and voluntary (*endogenous*) attention, the one driven by external stimuli, the other goal-directed and generated in brains, answers well to the facts that some stimuli "pop out" for particular animals more than others, and that many nonhuman animals seem to be able to direct internally their attention. But it is evident that the dichotomy cannot be a sheer one. The power of one external stimulus to capture attention while another does not has to do not only with the nature of the stimuli but also with organismal biases, which can range from innate dispositions of receptor cells and organs to an animal's memory of recent selective attention and rewards associated with it (Knudsen 2007; Awh, Belopolsky, and Theeuwes 2012). Attention, even involuntarily captured, is always the product of organismal capacities as well as information impinging on them. Attention involuntarily captured, also, often turns quickly into voluntary attention, as when I direct my gaze to locate in the bushes the blackbird I've heard, or engage my long term memory of features of blackbirds I've learned in earlier experience, comparing, for example, today's song with yesterday's. In addition, highly developed forms of attention in humans and some nonhuman animals can also involve the integration of features of stimuli into a broader target of attention. This can take stock of the context of a target, merging various stimuli — the movement of the blackbird that I have visually located as well as its overheard song.

Any simple separation of voluntary (top-down) and involuntary (bottom-up) systems is inadequate, then; in fact the joining of the two, involving multiple neural systems processing stimuli in reciprocal and recurrent connection, has yielded the best models we have for highly developed attentional systems. Eric Knudsen (2007) connects several elements in his multicomponent model for visual attention in vertebrates, a model generally adaptable also to auditory and other modes of attention. Here, in summary form, is how his model works:

1. Sensory neurons are innately attuned to different features of external stimuli and differentially activated by them. Thus they form *salience filters*, a first level of gatekeeping, transmitting selected information for higher-up processing. A honeybee's attention can be directed by a stimulus that begins with the sensitivity of a single neuron to sucrose concentration. My attention to the blackbird's song is one instance of an attentiveness starting from auditory input, dispersed across countless animals.

2. Higher-up processing, far from sensory receptors, engages *working memory*, the short-term storage necessary for evaluating incoming information and making decisions about it (Baddeley 2003). Such memory is present in many animals, including some invertebrates, but the complexity of its operation in birds and mammals outstrips that in other animals.

3. In these groups, at least, information that has made it past the first filters does not freely enter into working memory but is subject to further filtering, so that only the most important strains of information, relevant to a present circumstance, move up the processing chain. This *second filtering system* is largely controlled from the top down, through feedback from working memory, which modulates sensitivity to different kinds of informational input.

4. The modulation also involves the connection of working memory to other aspects of cognition, especially *long-term memory* and learning connected to it, and in the most complex attention systems it accesses *episodic memory*, a kind of long-term memory we will return to. This is how my experience of other singing blackbirds, in whatever richness I can remember, shapes my attentiveness to the present one.

5. The top-down feedback from working memory reaches not only to the level of the second gatekeeping or filtering, but also to *motor responses* to the world — for example, in the case of visual attention, gaze control: my searching to see the blackbird I've heard.

In this whole system, working memory exercises a crucial but not absolute control on what engages attention. Attention of this highly developed kind is not merely a neuronal salience assigned to certain stimuli, such as we find in honeybees, and neither is it simply deployed according to the dictates of executive controls. Instead, it is "an ongoing competition among information processing hierarchies vying for access to working memory. What is 'deployed' are top-down bias signals based on decisions made in working memory" (Knudsen 2007, p. 73). These biases are shaped also by the interactions of working memory with long-term memory.

Selective attention, in sum, depends on many components in interaction, including the attuned sensory filters, the neuronal pathways projecting to working memory circuits, the filtering effects of these, and the connections of working memory to long-term memory. It is generated through hierarchic processing across multiple levels linked in reciprocal relations. The neural substrates for its components, involving many brain *nuclei,* or bundles of neurons, have become clearer in recent years, and this has cast a bright light on differences in attentional structures between vertebrates and other animals.

Sensory receptors stand at one end of the attentional process, exercising their initial filtering functions. At the other end, far downstream from sensory neurons, are the centers for working memory in the cerebral cortex or pallium, connected to other centers where learned information is stored in long term memory. In between, deep beneath the cortex in the forebrain, lies the thalamus, the filtering interface between sensory input and memory and a powerful factor in overall synchronization of brain activities (Halassa and Kastner

2017; Knudsen 2018; Nani et al. 2019; Gu et al. 2020). In its mediating function it receives sensory information, relays it to the cortex, and sends impulses to motor neurons controlling responses. At the same time the thalamus is reciprocally controlled by cortical input, which acts selectively to suppress certain information it receives from sensory input. A separate nucleus surrounding the thalamus, the thalamic reticular nucleus, plays a role in this inhibitory regulation (Gu et al. 2020). It receives input from the cortex and the thalamus but, unlike the thalamus, does not send output to the cortex, so it can mediate between thalamus and cortex only in the direction of downward control.

These circuits extending from the thalamus to the cortex and back again — thalamocortical circuits — are found in highly elaborated form in both mammalian and avian brains, and in less developed form in other vertebrates: reptiles, amphibians, and fish (Knudsen 2018). There is no equivalent to them in invertebrate brains, despite the capacities these sponsor for integration of sensory information, memory, and associative learning (see Section 23). These brains can generate impressive effects, but they arise from simpler systems involving fewer distinct nuclei, fewer hierarchic levels, and fewer neurons overall.

In birds and mammals, the feedback circuits and hierarchies of processing produce a capacity for focus that extends far beyond a mere registering of salience. They bring about a distinct kind of focus in which salient stimuli are integrated with other stimuli and brought into a unified neural composite, joining multiple strains of informational input and forming a coherent moment of lived experience. Additionally, this complex processing places the integrated input in relation with other aspects of cognition, especially long-term memories, with their learned stores of data about the world. Altogether this results in a kind of soft-focus focus, an overdetermined attention in which a context for the focal target is established from other aspects of present experience and taken stock of or evaluated in relation to aspects of past experience. Such attention constructs a situation or episode in an animal's life. A songbird hearing the song of a conspecific places it in a context involving status, kinship, location and territorial occupation, access to mates and, as we will see in Part III, additional situational nuances.

Situational attention brings further consequences. Its contextual nature divides up the perceived world in a way that is not merely discriminative, separating one bit of input from another, but instead a constructed, hierarchized perception of part to whole, of focal target to its broader frame. The focus is at once integrated into something larger and distinguished from it — a working definition of *analysis*. If a niche comprises the external data coming to an animal from moment to moment, situational attention structures the niche as a nested hierarchy in which phenomena are grouped in larger coherencies that can be accessed through a single, phenomenal point of entry. Such analysis involves recursive parsing of the world, since what is a target of attention in one context can become, within the functional range of an animal's sensory and cognitive apparatus (for example, its attentional filters), a context for a more sharply focused target. A songbird can focus on another bird's song as a gesture lodged in a social context. In many circumstances it can also focus more narrowly on an element within the song in relation to the whole song — a gesture within a gesture. Advanced attentional capacities are inherently both situational and recursive.

What is described here is a distinct type of perception of the world, brought about by multiplex, hierarchic, and reciprocal neural processing. From a semiotic vantage it is also something more: a foundation for indexes, the signs nonhuman animals construct. It is not in itself semiotic, and there is in it no inherent relation to a relation. But the possibility of indexicality hovers close by, prepared in two ways. Cognition of hierarchized relations of part to whole is a requirement for the perception of the partial, aspectual relation between sign vehicle and object — the defining metarelation of all semiosis. And, as we have seen, the formation of an indexical interpretant starts from the coherence of a full situation, remembered and learned holistically but at the same time parsed into its elements.

Thus attention of a particular sophisticated and situational kind provides the foundation for semiosis and meaning. But signs require an additional building block, formed from the remembrance of whole, coherent life episodes and the learning it fosters. Brought into relation with the integrated data of novel situations, this episodic memory founds indexical semiosis.

FOURTEEN

Episodic Memory:

Assembling the Index

The working memory central to complex attention is narrowly con-strained in both duration and holding capacity, unlike long-term mem-ory in both these respects. These are not limitations so much as features shaped to its function, which combines temporary storage with fluid processing and manipulation of information, so enabling evaluation of moment-to-moment events and planning of responses to them (Bad-deley 2000). Like attention in general, working memory has its own multicomponent, multilevel structure, according to several models of it. These have mainly been devised and tested in humans, but some-thing like working memory plays a fundamental role in regulating the attention and behaviors of many animals, including invertebrates as well as vertebrates. (We will examine honeybee memory in Part III.)

The relation of working memory to long-term memory is an addi-tional complication — yet another network of components and systems of components — that has come to be considered essential to the full integrative powers of advanced, situational attention. For Alan Bad-deley, originator of one of the main models of working memory, long-term memory forms a "crystallized" system of contents that enables, in the humans he studies, the formation of remembered life episodes (Baddeley 2003). Bringing these into interaction with working mem-ory enables the construction of the thick contexts necessary for com-plex attention, planning, and response. It requires that the episodes of long-term memory be rendered fluid in the interaction, and Baddeley posits an *episodic buffer* as the neural component that achieves this — a limited-capacity system that retrieves episodes from long-term memory storage and integrates them with the transitory information in working memory. The episodic buffer mediates between the momentary flux of attention to ongoing events, with its mobile executive processing of working memory, and the fixed memories of past episodes, shaping the one according to the other. Behind situational attention, then, connected to its working memory through an episodic buffer, stands a memorized store of life episodes. What is this store? How does it arise, and in what neural substrates can we locate it? How far beyond humans does it extend in the animal world?

The term *episodic memory* was introduced by psychologist Endel Tulving (1972) to name the kind of memory in humans that lodges events or objects in a broader context. It is also referred to as what-where-when memory, since it often includes both a spatial and temporal location for its contents. For Tulving, an additional and uniquely human feature of episodic memory is its *autonoetic* or autobiographical quality, which separates it from semantic memories encoding learned facts disconnected from explicit situations (Tulving 2002). I can remember, for example, the blackbird I heard singing yesterday morning and then spotted in the bushes — an episode in my life — and I can also remember that the blackbirds I've seen in Europe have yellow beaks, while those in my US locale have red shoulders — learned facts retrieved without episodic contexts. Humans have immense capacity for both kinds of long-term memory. Their remembered episodes can vary widely in richness of nuance and detail, regularly extending far beyond a simple what-where-when matrix. Episodic memory can bring about episodic learning also, a contextual learning that shapes ongoing behavior according to coherent remembered episodes — for example, my scanning today the bushes where I saw the blackbird yesterday in the hope of seeing it in the same place again (Nuxoll 2012).

The anthropocentric focus of Tulving's episodic memory began to be widened in the 1990s, when experiments by Nicola Clayton and

Anthony Dickinson on scrub jays showed that they employ episodic memories in their food caching and retrieval (Clayton and Dickinson 1998; Salwiczek, Watanabe, and Clayton 2010). Acknowledging the unknowable or untestable nature of a bird's autonoesis—the inaccessibility to us of what it is like to be a bird—they cautiously termed their bird memories *episodic-like*. More recent research on further birds, rodents, nonhuman primates, and other mammals has solidly established the existence in many nonhuman animals of memories of things and events in context, encouraging the application of episodic to all these groups (Templer and Hampton 2013). Meanwhile, neurobiological studies have pinpointed key components in the neural architecture involved in human episodic memory, and other animals capable of episodic memory have been found to possess structures that are homologous—connected in evolutionary descent—or, in the case of birds, partly homologous, partly analogous—arrived at through both descent and convergent evolution. The neural substrates connected to episodic memory, ample evidence now indicates, are highly developed in mammals and birds, less developed in other vertebrates, and absent in other animals (Allen and Fortin 2013).

Like analytic, situational attention, episodic memory depends on hierarchic processing across a variety of interconnected neural areas, some neocortical and others deeper in the brain. The major components of this network in mammals, according to Timothy Allen and Norbert Fortin, are integrative areas in the prefrontal cortex, the deep-brain hippocampus, and several parahippocampal nuclei mediating between the two. Structures closely related to all three are found also in bird brains (Allen and Fortin 2013). The prefrontal cortex receives sensory data from other cortical areas, integrating it across sensory modalities, thus associating, for example, my hearing and seeing of the blackbird. The multimodal complex resulting from this associative processing is relayed through parahippocampal nuclei to the hippocampus. This laminated, multiplex structure is strongly implicated in the formation of spatial memories in many vertebrates, and "place cells" firing in a fashion attuned to specific locations in an animal's environment are well documented in the hippocampi of rodents. This suggests an original function of the hippocampus in vertebrate navigation as well as the original form of the "where" aspect of episodic memories.

In mammals and birds, however, the complexity of memories associated with the hippocampus reaches well beyond spatial location alone to sponsor what-where-when memories of varying richness. Unlike the thalamus involved in the formation of situational attention, the hippocampus receives no direct input from sensory receptors but instead functions to process further and store the integrated complexes it receives from the prefrontal cortex. It is probably responsible for the full integration of this complex into a remembered episode, in whatever degree an animal achieves this. "Episodic recall," then, "is thought to occur when the integrated event-in-context representation is reactivated"; it takes the form of "a pattern completion process," and so requires, to be set in motion, only elements or aspects of the context, acting as cues to the whole context stored in the hippocampus (Allen and Fortin 2013, p. 10383).

The cueing — to put together, now, the attentional and the episodic memory networks — involves interchange between the episodes stored in long-term memory and new sensory data brought to salience and processed in working memory. This interchange is what Baddeley aims at with his idea of an episodic buffer, functioning both to retrieve memories and bind them to new, salient information filtered through the attention system. From this binding emerges the evaluation of and response to new experience, and the model suggests that the richness and nuance of these might be proportionate to the richness of episodic memory. Overall, the conjoined networks of attention — from sensory receptors to neocortex, with the thalamus between — and episodic memory — from prefrontal integration to hippocampus and back to the centers of working memory — make up the neural foundation for situational awareness, evaluation, and response in the behaviors of some animals.

We can now assemble the ingredients that lead to semiosis. We have seen that the capacity for situational attention forms a foundation for indexicality, and the examination of episodic memory reveals its own contribution to this foundation. In providing the episodes that enable situational attention, it creates the kind of learning the index relies on, the interrelation of present data and past situations that fosters the indexical interpretant. Its contextualism brings to cognition the part-to-whole hierarchy that enables analysis, and its episodic singularity, distinct from the modes of abstraction that make both iconism and symbolism possible, founds the singularity of every index. In the congruency of episodic memory to indexicality, we find an explanation for the prevalence of indexicality among non-human animals; we also find good reason to propose that this indexicality extends only as far in the animal world as episodic memory and the situational attention to which it is so closely related.

This is tantamount to charting the limits of semiosis all told; and as goes semiosis, so goes meaning. The modeling of the cognitive systems for situational attention and episodic memory, the identification of the neural substrates responsible for these, and the tracking of the limits of these substrates together provide circumstantial evidence to locate the border of sign-making and meaning in the animal world. We see emerging here, finally, the conditions that define the fourth abstract machine, the machine of semiosis:

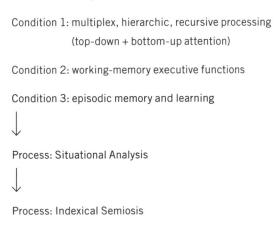

Condition 1: multiplex, hierarchic, recursive processing
(top-down + bottom-up attention)

Condition 2: working-memory executive functions

Condition 3: episodic memory and learning

↓

Process: Situational Analysis

↓

Process: Indexical Semiosis

FIFTEEN

Principles

To review the argument so far, here are some guiding principles for the attempt to locate meaning, with their corollaries:

The ubiquity of informational process: *There is a continuous spectrum of biotic informational processes reaching from almost instantaneous molecular interactions to long evolutionary* durées.
- Categorical distinctions between short and long processes are artificial or heuristic.
- Analysis of a process can range across the whole spectrum of temporal levels it involves.

Mediation of organismal action: *All behaviors arise in nonlinear networks of informational mediation, which form dispositives extending from intracellular molecular regulatory systems through organisms, groups, and superorganisms to ecosystems.*
- Mediation, like information, is ubiquitous.
- No organismal action or effect is directly determined by genes.

Embedded mediation: *All organismal actions are embedded processes generated in covariance with surrounding conditions, that is, with an environment in some degree altered by the organism in it, a niche in some degree constructed.*
- All content, meaning, mindfulness, and their byproducts (culture, technology, and more) are products of niche-constructive mediation.

Information without content: *All information is causal covariance and need not convey content.*

- In most life-forms, all processes at work involve information without content.
- In all life-forms, most processes at work involve information without content.
- Most biotic information is meaningless.

Complexity from causal information: *No content is necessary for organisms, including animals, to generate complex behavior.*

- In the face of behavioral complexity, the presence of content is not a default postulate but needs to be supported by identification of processes that create it.

Information with content: *Information with content, semantics, reference, meaning, or aboutness is a special subset of causal information.*

- In some life-forms, some processes at work involve semantic information.
- All life-forms fall into one of two types, causal-informational and causal + semantic-informational.
- Semantic information supervenes on causal information.

Semantic information from signs: *Signs are the fundamental units of content.*

- Meaning arises from semiosis, the production of signs.

Mediation of meaning: *Signs involve processes of informational mediation distinct from those in causal information.*

- Semiotic mediation is defined by processes creating the ingredients of semiosis, including interpretant and metarelation.
- Semiotic organisms all share general features of their information processing not found in nonsemiotic organisms.

Types of signs: *Semiotic mediation results in types of signs differing in the relations in them of sign, object, and interpretant, which are dependent on the capacities of the animals creating them.*

- In nonhuman semiosis in the world today, the index is prevalent, while icons and symbols are rare or nonexistent.

PART THREE

Meaningful and Meaningless

Complexity

Among the intricate lifeways of animals, how can we differentiate

those that exploit the semiotic machine to make meaning

from those that process meaningless information?

In the Realm of Aboutness:

Songbirds

SIXTEEN

Birdsong Basics

Half a century's intensive study of birdsong, involving hundreds of researchers from many fields and many, many thousands of birds, has yielded a broad-based account of the phenomenon (for overviews, see Marler and Slabbekoorn 2004 and Catchpole and Slater 2008). Here in summary form is what we've learned.

Almost all birds vocalize as an aspect of their social interactions. In the majority of species the vocalizations take the form of innate calls: fixed, unlearned, and not very varied in their phonological structure. In thousands of other species, however, something differ-ent has evolved: vocal learning, the ability to master during ontog-eny a repertory of species-specific sounds subsequently employed in social interactions. Vocal learning is concentrated in three taxo-nomic orders: parrots (order Psittaciformes), hummingbirds (order Apodiformes, family Trochilidae), and *oscine passerines* or songbirds proper, members of the order Passeriformes, which includes also *suboscine*, non-vocal-learning species. It is an open question whether these groups share a common vocal-learning ancestor or converged independently on the capacity. The favored view sees hummingbirds as an independent, convergent line, with parrots and songbirds more likely to have shared an ancestral vocal learner (Berwick et al. 2012; Hara et al. 2012, fig. 1).

The number of oscine passerine species is usually estimated at about five thousand, though this number can change radically according to taxonomic protocols—another reminder of how difficult species are to define and identify. Whatever the number settled upon, songbirds are the largest group of vocal learners. Parrot and hummingbird species number about four hundred each, and apart from birds vocal learning is rare, found only in a small number of mammals: many cetaceans, some pinnipeds, bats, elephants, perhaps mice and goats, and *Homo sapiens*, alone among living primates. In other mammals, as in non-vocal-learning birds and many other animals besides, vocal communication is restricted to unlearned, innate calls. Divergences from this consensus judgment are certainly possible, and the roster of vocal learners might be expanded. This is because vocal learning is best understood as existing on a continuum, gauged according to the measures of several capacities involved in it, with no sharp border between its presence and absence (Nowicki and Searcy 2014; Martins and Boeckx 2020). Moreover, some birdcalls move toward songs on this spectrum and involve some degree of learning (for one example, the chickadee, see McMillan et al. 2017). But whatever criteria we use and whatever list results, highly developed vocal learning is dispersed among songbirds on a scale that dwarfs its appearance in all other animals.

In oscine passerines, songs are typically learned during a sensitive period in juvenile development, though there are many species of open-ended learners where song remains malleable and learning continues into later life. Juveniles pass through two main learning stages, a phase of auditory acquisition of species-specific song and a later phase of production of it. The first phase has been thought to lay down a memorized version of the song to be reproduced, though recent evidence has complicated this "auditory template model," since it is now known that flexibilities built into the shaping of song can last well beyond the memorization phase even in species that are not open-ended learners. The second phase, a production or motor phase, involves a practice loop in which a bird listens to itself producing song or its antecedent elements, repeats them, and more and more closely approximates the model song. This phase normally moves from a period of *subsong*, often likened to the babbling of human infants, through plastic versions of the song or its parts, to the crystallized final version or versions. Open-ended learners sometimes revert at the beginning of each new season to something like the sub- or plastic song stage (Mori, Liu, and Wada 2018). Learning species-specific song involves an innate component but also depends on listening to nearby adult conspecifics (Belzner et al. 2009). These tutors are usually unrelated to the learner — not supportive teachers so much as potential competitors.

Birdsong is combinatorial, constructing songs from a set of small, learned gestures or syllables. Analysts further subdivide particularly complex syllables into smaller units (notes), and there is growing evidence that many species distinguish still finer nuances within notes that human observers have tended to consider identical (McMillan et al. 2017; Fishbein et al. 2019). Syllables, in any case, are important units, not merely an artifact of human analysis. We know this because songbirds in the learning stage show neuronal firings in relevant brain regions at the onset of both their own babbled syllables and the syllables they hear from their tutors. These neurons can be selective for syllables with certain phonological features, suggesting a mechanism militating toward species specificity in the resulting songs and a similarity to the "perceptual magnet effect" by which human infants are channeled toward the phonological categories of the language they hear (Marler and Peters 1977; Ribeiro et al. 1998; Mackevicius, Happ, and Fee 2020; Guenther and Gjaja 1996).

From notes and syllables are built motifs or phrases, each comprising several syllables, and from these are built whole songs. Species differ widely in the number of syllables they learn and the length and complexity of the songs they build from them, but the syllables are basic to the species-specific identity of the songs. Complete songs in most species are repeatable wholes — song types — recognizable by their syllable content, syllable order, and timing. Species show repertories ranging from a single song type repeated with minimal variation (the black-capped chickadee) or a single type with variation (the zebra finch is a much studied example) to dozens (European blackbird) or hundreds of types (canaries, marsh wrens); most species are intermediate between these extremes (cardinal, song sparrow, 7–12 types). Some species seem not to form discernible repertories at all but instead recombine their syllables into long songs without ever exactly repeating a type (sedge warbler). (For approximations of some repertory sizes see Catchpole and Slater 2008, p. 205.) Many birdsongs manifest also what can be thought of as syntax — varying probabilities of certain syllables or motifs following certain other ones in the combinatorics of a song (for Bengalese finch syntax, see Okanoya 2002; Berwick et al. 2011; Lachlan et al. 2010). Such syntax can extend to the level of whole songs, with certain song types following certain other ones in the unfolding of an extended song bout.

It is a commonplace that the songs of oscine passerines are associated with two important and related areas in bird sociality, mate selection and territorial assertion and defense. These are broad, general categories, involving many distinct behaviors and their innumerable variations among thousands of species. So general are they, indeed, that the attempt to match them with the dazzling variety of song and its situational deployment can obscure the nuances we need to understand (Catchpole and Slater 2008, pp. 236–39). The problem is an instance of the catch-all, adaptationist assignment of evolved function that we examined in Part II. Phenomena such as territoriality and mate selection do not identify functions so much as they suggest trends we might expect to find in the wide realm of song practice.

The association with sexual selection suggests that sexual dimorphism in the production and use of song might be frequent, and many of the most studied species, which have been native especially to northern temperate zones, show this clearly, with learned vocalizations largely or wholly restricted to males. (Even in these species both genders vocalize innate calls, not subject to learning.) In many of these song-dimorphic species, general correlations are well established between male song performance and female mate preference, expressed in changes in behavior and physiology. In some cases other, more specific correlations have been detected between mating success and the length and complexity of song types, the size of song repertories, and the overall output of song (for a review, see Catchpole and Slater 2008, chap. 7). Analyses of these correlations have examined the metabolic costs involved in song learning and production, which as they rise might vouch for the "honesty" of male singing as a signal of mate quality. Recent research has also modeled the impact of repertory size on niche construction more generally, including breeding success and the ontogeny of song learning (Creanza, Fogarty, and Feldman 2016). Song learning itself, moreover, can take multiple forms even in strongly dimorphic species, extending beyond learned production in one sex to learned responses to song in the other. It has been shown in some dimorphic species that, while males rarely learn song from their fathers, females often learn to prefer songs close to those of their fathers (Catchpole and Slater 2008; Bolhuis and Gahr 2006). In Case 3 in Section 19 we will examine an instance of particular "sexy syllables" sung by male canaries and their effects on potential mates.

In the midst of this emphasis among researchers on sexual selection and dimorphic species, the wide prevalence of female birdsong has only gradually become clear, especially as more and more tropical species have been studied, where dimorphisms are less pronounced and typical (Riebel et al. 2019). In fact, phylogenetic studies now reveal that song shared between the sexes is the likely ancestral situation, with dimorphism emerging later among many species (Odom et al. 2014; Price 2015). There remain large gaps in our knowledge of female song, to the extent that, as a recent review notes, in only about a quarter of all songbird species can we say with confidence whether females sing or not. What is clear, the review continues, is that "Across songbirds, there is considerably more variation in female than male song, ranging from species with no female song to species in which female song output or repertoire sizes are equivalent to or more extensive than in males" (Riebel et al. 2019, pp. 2, 4).

The association of song with territorial defense likewise suggests a trend evident in many species of northern, nontropical songbirds:

song as a regulator of male-male competition (King, West, and White 2002; Beecher and Burt 2004; Catchpole and Slater 2008). This often influences juveniles' choice of tutors, where powerful and even threatening adults can be preferred; in the so-called eavesdropping model of tutelage, young birds listen to the song interactions of adult male rivals, choosing to learn from the dominant bird (Beecher et al. 2007). Between adult males, song competition can take subtle forms involving portions of one male bird's repertory — several of its song types — that are shared with a neighboring male in "matched countersinging" between the two (Beecher and Stowe 2005; see Case 1 in Section 18). Here again, however, a broadened canvas reveals an array of possibilities for territorial marking or defense other than two males competing in song. These include duetting, usually between female-male pairs, common in the tropics and subtropics (Case 2, Section 18), and chorusing among groups of birds, which can range acoustically from precise alternations of phrases sung by two groups of birds, female and male, to much freer, even cacophonous sound-assemblages. (For examples of all these, see Catchpole and Slater 2008, pp. 215–25; for granular structure of duets, Hoffmann et al. 2019.) In pair-bonded or communal species, such deployments of song seem often to be involved in joint guarding of territory and brood, but the nuances of possibility here are many.

In the 1970s researchers began to isolate a neural infrastructure that subserves the learning and production of birdsong (Nottebohm, Stokes, and Leonard 1976). Since then hundreds of studies have been devoted to detailing the nature and working of this *song system* or *song control system*. The system seems to be common to oscine passerines (songbirds proper), and recent investigations have shown similar systems in the other vocal-learning birds, parrots and hummingbirds, if with differences in the particulars of network circuitry (Gahr 2000; Chakraborty et al. 2015). There is evidence of rudimentary versions of some of the system's structures in the brains of suboscine passerines, suggesting a connection of the songbird system to less hypertrophied brain networks in these non-vocal-learning animals (Liu et al. 2013).

We will return to the song system. For now, it is enough to caution that it should not be thought of as an autonomous, static structure determining the sung sociality of birds that possess it but instead as a plastic set of networks in songbirds' brains involved in an ongoing interaction with their lifeways and niches. Its plasticity is evident at temporal spans ranging from a lifetime through seasonal rhythms down to much quicker, even momentary changes. In processing auditory input and instigating sung output it engages multiple levels of mediation that reach from social to molecular scales, where regulatory molecules such as immediate early genes and neurosteroids, stimulated by sensory input, alter within minutes the expression of other genes, altering also neuronal transmission in the bird's brains and bringing about corresponding changes in their bodies (Jarvis 2004; Wang et al. 2019). The song system, in other words, is a process involving neurons, neuronal nuclei, networks of nuclei, general brain architecture, body, other birds, and the niche. All its components — right down to the genetic information involved — are malleable elements engaged in the niche-constructive feedback loops of songbirds.

SEVENTEEN

Birdsong Meanings

and Complexity

More precisely, the song system is a *semiotic* process — an operation, widely dispersed, complex, and staggeringly varied, of the abstract machine described at the end of Part II. The following sections will show how this is so. General semiotic consideration of birdsong is rare, even though it offers the advantage of directing our understanding toward the underlying process by which its meanings emerge. Nevertheless, the fact that the songs are meaningful tends to be taken for granted by most researchers, acknowledged or implied usually at the level of the generalizations about territorial assertion and sexual interactions described before: One song says, "I'm here: get off my yard," while another asks, "Won't you have me as a mate?"

Focused inquiry on birdsong meanings, rather than taking a transspecies semiotic approach, has since the end of the twentieth century tended to compare the songs with human language. It is an almost irresistible strategy, given the many general features shared by the two phenomena: learning concentrated in a juvenile sensitive period, babbling behavior during early learning, combinatorial construction, hierarchic syntactic structures, and — obvious enough to be overlooked — phonologically complex and varied vocalism. Resistible or not, the comparison of birdsong and language needs to be pursued cautiously, prone as it is to a particular version of humanist parochialism that sets human language as the high bar against which other communicative phenomena are gauged and understood. I have called a similar tendency in human evolution studies *linguocentrism*, where language is mistakenly presumed to be a prerequisite for the emergence of nonlinguistic behaviors such as musicking and certain aspects of ritual (see Tomlinson 2015, 2018). Human language research no doubt needs birdsong research, since it can tackle experimentally issues difficult of access or entirely inaccessible in human infants (for an instance of well-judged comparativism, see Berwick et al. 2012). But it's not clear that the converse is true — that birdsong research needs human language research — and in every case where human capacities serve as a benchmark, care needs to be taken not to trample avian difference underfoot.

Language/birdsong comparison in recent decades has gravitated toward the generative, formalized grammar of Noam Chomsky, which has kept the relation of syntax, lexicon, and meaning at the forefront of linguistics since it was introduced in the 1950s. This work has proposed a categorical difference between the combinatorial structuring of birdsong and language that revolves around the levels at which meaning can be found. Meaning in human language arises from words or parts of words (for example, *sub-* or *–s* in "suboscines") that carry significance, arranged according to syntactic requirements for well-formed sentences. Meaning inheres in both levels, the bits and the larger assemblages of them — a semantic duality that manifests a particular form of combinatoriality that linguists term *compositionality* (Bolhuis et al. 2018, p. 4; Bowling and Fitch 2018). Words fall into categories (that is, parts of speech) that are constrained in particular ways by syntactic rules, and reordering words within the constraints can shift the meanings that result: "Juliet rescued a bird." vs. "A bird rescued Juliet."

Generative grammar details how this dual construction of meaning allows for complexities in human language such as serial dependencies ("Juliet brought [the injured bird] [in her car] [to the vet]."), nested dependencies ("The bird [Juliet found in the yard] was injured."), and recursion, by which full grammatical categories can be nested within other categories of the same type (for example, sentences as parts of other sentences: Serial: "Juliet was sure [the bird injured itself by flying into the window]." Nested: "Juliet related how [the bird injured itself] to her friend." Multiple: "Juliet thought [the bird flew into the window] because [the cat surprised it] while [it was eating the worm]."). From Chomsky's work have been developed several algorithmic models, commonly thought of as computational automata, arranged in levels of increasing power necessary to produce all this syntactic complexity. In this "Chomsky hierarchy," human language requires a *pushdown automaton*, equipped with memory retrieval capacity that can peel off memories one by one to reveal successively deeper ones, and, more specifically, a complex version of this, a *nested pushdown automaton*, with a hierarchy of sub-banks of memories accessible through memories exposed in the primary memory bank (Balari and Lorenzo 2013). Well-developed working memory and episodic, situational memory, of the sorts discussed at the end of Part II, are modeled in these automata and are necessary for constructing the dependencies that characterize language.

Unlike human language, birdsong seems to have no lexicon of meaningful particles that enter into its syntax. Its notes, syllables, and motifs are not equivalent to words, and its syntax arranges units that are acoustically but not semantically distinct — a fact that has led it to be characterized as phonological syntax, contrasting with the semantic or lexical syntax of language (Marler 2000; Berwick et al. 2011; Berwick et al. 2012). There have been several recent interpretations of experimental data arguing for modest particulate meanings of individual notes or syllables and for the kinds of syntactic complexity they can generate (Abe and Watanabe 2011; Engesser et al. 2015; Suzuki, Wheatcroft, and Geisser 2016), but in each case these have been justly criticized for overreaching the data (Beckers et al. 2012; Bowling and Fitch 2015; Bolhuis et al. 2018). The experiments they report demonstrate intriguing complexities of birdsong syntax and suggest even that overall meanings of songs might be modulated or underscored by shifts in the nature or frequency of particular syllables or motifs — we return to this issue in Case 3 — but they do not support a full-fledged, separate level of particulate meaning analogous to words. Birdsong combinatoriality is not compositional in linguists' sense of the word.

Because of this nonsemantic nature of their basic combinatorial units, birdsongs do not construct the dependencies evident in linguistic meaning and exemplified earlier. Their syntax does not rely on the immediate access to changing memory input afforded by pushdown automata, needed for such dependencies, but is describable using a simpler grammar formalized in the Chomsky hierarchy. This is modeled by the *finite state automaton*, in which each state of a system (that is, each note, syllable, or motif reached in a birdsong) can be followed by a transition only to one of a limited or finite array of new states (other notes, syllables, or motifs). This is a version of a Markov chain, in which the probability of each new state (motif or syllable) is determined only by the one that immediately precedes it (Okanoya 2002; Berwick et al. 2011; Berwick et al. 2012). Birdsong operates mostly — perhaps entirely — within finite state limitations on dependency structures. In human language, instead, such constraints are superseded by the operation of pushdown automata — that is, more complex and hierarchized memory capacities.

This comparison of birdsong and language according to formalized grammar highlights a significant difference between the two, but we can see that the questions it raises are limited by its focus on generative syntax and that the conclusions it reaches tend to be couched in negative terms, registering absences at the heart of birdsong. The "hierarchical depth" of birdsong structures is pronounced to be "strictly limited," and the meaning generated is judged to admit no "compositional semantic creativity" and to convey only the most basic "conceptual-intentional component" (Berwick et al. 2011, pp. 115, 118; see also Berwick et al. 2012). Birdsong is in several aspects probably the most complex of all nonhuman communication systems, yet, viewed in the long shadow of language, it is diminished to a stumbling aspirant to human attainments.

Here the burden of linguocentrism weighs heavy, in spite of researchers' awareness that even a finite state machine can generate structures of imposing complexity. In birdsong these are realized in several ways. Hierarchies of construction, where finite-state syntaxes are activated, can operate on several levels, from note to syllable to motif, and the phonological units joined syntactically in larger wholes can come from any of these levels — the unit can take the form of a note, notes joined in a syllable, or syllables joined in a motif. The syntax of many song types, moreover, allows loops backward in the sequence, enabling the flexible expansion of these songs through repetition or variation of sequences, which can similarly use units at several hierarchic levels. Looping can result in structures that resemble the nested embedding of language, with certain motifs, syllables, or notes lodged within repeating occurrences of others. All these are indicated in a finite-state diagram of a song of a Bengalese finch (see Figure 17.1). These are rule-governed syntaxes, but phonological ones, without semantic dependencies or parts of speech.

This structural complexity poses the broadest puzzle of all about birdsong, one that recurs like a perplexed watchword across even authoritative surveys (see, for example, Catchpole and Slater 2008).

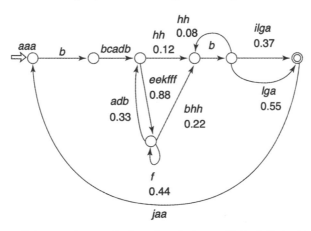

Figure 17.1. Finite-state diagram of a Bengalese finch song. Circles indicate states, letters notes, and numbers probabilities of particular transitions, where more than one state transition is possible. Reprinted from Robert C. Berwick, Kazuo Okanoya, Gabriel J. L. Beckers, and Johan J. Bolhuis, "Songs to Syntax: The Linguistics of Birdsong," *Trends in Cognitive Sciences* 15 (2011), p. 117, with permission from Elsevier.

If it's just a matter of attracting mates or defending territory, why the huge diversity, richness, and complexity of birdsong practices? Neither functional correlation at the general level of territory and sex nor comparisons of birdsong syntax with language can explain the hierarchized internal complexity of individual song types, with their varied combinatorial possibilities. They cannot illuminate the presence of song repertories in the majority of songbird species — not even modest repertories of five, eight, or twelve songs, let alone the immense repertories of certain species, with dozens or hundreds of song types. To maintain these behaviors requires expensive wetware mechanisms and considerable energy devoted to learning, memorization, and production. Adaptationist and linguocentric approaches hardly begin to reveal what justifies the expense.

Across a long evolutionary history involving thousands of species, of course, diverse levels of complexity might be predicted in both whole, species-specific repertories and individual song types within them. It's implausible to think that selection and niche construction would not conduce to variety and, in some instances, hypertrophy of the capacities at stake. Understanding these cases requires a holistic view of individual species' lifeways, examining altogether the niche-constructive assemblages of behaviors and affordances they involve: processes involving innate but plastic organismal capacities, sensory input from environments, responses to it, and mutual alterations of organism and niche. Such a view predicts abundant variety in the ways songs unfold in feedback processes connecting them with other aspects in the assemblage, which in turn predicts variety in the ways songs achieve their effects, with possibilities shifting in tandem with the unique niche of each species and constrained only loosely by the song-producing, brain-and-body infrastructure common to songbirds. This view also predicts the plasticity of the infrastructure — its operation, again, more as a process than as a static architecture — as it is continually shaped by the sociality and habitation of living birds.

All this also describes, within the semiotic process, the formation of interpretants.

EIGHTEEN

Avian Interpretants

Let us quickly review some semiotic conclusions we reached at the end of Part II. The niche of an organism is an assemblage of features of the external world that rise over its affordance horizon. For a semiotic organism, some of these features function as sign vehicles and some as objects in the sign relation. This relation is constrained from both sides. The objects limit, according to their natures as perceived by the organism, what particular aspects of other phenomena might enter into a signifying relation with them. Those phenomena, meanwhile, enter into the sign-relation — become sign vehicles — under their own constraints, which variously isolate aspects of their objects according to three kinds of relation: relations of qualitative likeness (icons), singular processes and events in which both are involved (indexes), or generalizations or laws governing their deployment (symbols).

Signs require, in addition to this sign/object relation, perceptual and cognitive processes that isolate aspects of potential objects according to aspects of potential sign vehicles. This correspondence is the interpretant, in which animals register the aspectual relation of object to vehicle according to the constraints exercised by both. The capacities to form interpretants — advanced attentional focus and its recursive parsing of the world, episodic memory, and learning abilities associated with it — are prerequisites for semiosis, rare attributes distributed through one or several corners of the animal world. An interpretant manifests an animal's ability to construct a metarelation, establishing the relation of the percept it can form to another relation, of vehicle and object, and thus creating aboutness and meaning.

The process of animal semiosis is related to more general organismal processes in that it is always niche-constructive and always connected to the mediated, nonlinear networks linking the external world to the organism, down to the molecular level. Meanings need to be understood at once as processes of metarelational interpretant formation and as a special kind of niche construction. This can hardly be overstressed: Semiosis is not a process that takes place and then shapes a niche; neither is niche construction a process creating a habitat in which an animal then goes about its sign-making. The interpretant process instead links the two in a special kind of mutual development, determining the course of niche construction even as the niche constrains in myriad ways the signs it can generate. Such mutuality is universal in the relations of organisms and niches (we will linger over it in the case of bees), and semiosis is nothing but another instance of this, however exceptional.

The type of semiosis enacted in birdsongs, as in virtually all non-human semiosis, is indexical. The abstraction of pure iconism is not a part of avian semiosis, since, as we will see, their songs are pointers in broader, eventful social processes and episodes. Neither is there any set of laws or conventions determining the roles of birdsongs in bird lives or the differential structures of the birdsongs themselves, as symbolism would require. Avian indexicality is by no means a simple thing, however. The syntactic complexities of birdsong and the multiple repertories of most species point to something more — or many, many *somethings* more in different species and niches. They reveal incipient categorizations of semiotic gesture, formed at the levels of note, syllable, motif, and song type; arising from these, they show in some cases a hierarchization of the sign vehicles themselves. These together suggest a shift on the spectrum, not to full symbolism but toward the *hyperindexicality* introduced in Section 12.

Three case studies can clarify birds' indexicality and illuminate their formation of interpretants. In each one the birdsongs (or parts of them) are the sign vehicles, hereafter simply "signs." They are not the only signs that birds can produce, but they warrant special consideration because of their extraordinary nature in several dimensions: their learned development, hierarchized structures, richness of repertory, and flexible deployment in bird sociality. All these features are aspects of the interpretant process that creates the sung sign. All arise at the juncture of complex cognition and rich social niche construction.

Case 1

A male song sparrow in the northwest United States (*Melospiza melodia morphna*; Figure 18.1) learns a repertory of ten or so song types across a period that can extend through the bird's first summer and into the following spring. It learns from tutors near to where

Figure 18.1. Song sparrow (*Melospiza melodia morphna*), drawn by Virge Kask after a photograph by Tony Hisgett.

183

it will establish its own territory (Beecher and Burt 2004), with the result that repertories of song types overlap according to geographic proximity, with some percentage of a bird's songs shared with its neighbors, and a particularly large percentage shared with its primary (and neighboring) tutor. This relation seems to be beneficial to the tutor as well as the tutee, and a positive correlation has been detected between the proportion of its repertory a bird learns from its tutor and the length of survival of both on their adjacent territories: the more songs neighboring birds share, the longer (on average) they occupy their territories (Beecher, Çaglar, and Campbell 2020). Song sharing between these sparrows is probably best understood as a "dear enemy effect" of mutual tolerance rather than true cooperation or alliance, and it is distinct from two other models of song tutelage evidenced in other species: a competition model, in which a young bird enters into direct rivalry with its tutor, and an eavesdropping model, in which the tutee listens to interactions of adults, perhaps gauging which of them is dominant and choosing it as a tutor (Beecher, Çaglar, and Campbell 2020). Song sharing among sparrows forms an integral part of the general dynamic shaping their repertories. They sometimes trim their repertories in their first full spring, coordinating them more closely with those of their neighbors, and, given the similarities of the syllables making up the songs, even songs resembling one another by chance rather than tutelage can come to be treated by birds as shared (Beecher and Campbell 2005). All this has been revealed in an ongoing series of playback experiments in the wild by Michael Beecher, John Burt, Elizabeth Campbell, and their co-researchers.

Even between mutually tolerant neighbors, however, there exists the possibility of threat and aggression, and the experiments of these researchers have revealed an intricate register of differences based on song sharing that helps to manage coexistence and stabilize both territories (Beecher and Campbell 2005). This involves *countersinging* (Catchpole and Slater 2008) in which one sparrow directs toward a neighbor a song type shared between them. The neighbor can respond with the same song or a very similar one (a type match), with a different song shared between the two (a repertoire match), or with an unshared song. Type matching is an aggressive response and can elicit from the first bird an escalation, signaled by its own type-matched counterresponse. It might lead the first bird to approach the birds' shared border, engage in wing-waving displays, and sing a distinct kind of soft song, a prelude to outright conflict (Templeton et al. 2012). A repertoire match by the neighbor is less aggressive than a type match, an equivocal response that can elicit from the first sparrow a type match (escalation) or a different song (de-escalation). Responding with an unshared song is plain de-escalation or aggression avoidance. Long-term neighbors tend to respond with repertoire matches more than with type matches, while new neighbors are more prone to type matches, especially in their first breeding season in the neighborhood. Thus, unfamiliar birds are more likely to elicit from settled ones a response appropriate to a threat to their territories. In this way the songs help modulate relations in which long-term neighbors are distinguished from new and perhaps more "expansionist" ones (Beecher and Campbell 2005; Beecher et al. 1996).

The sung typology of shared/unshared songs and, in countersinging, the further differentiation of type matching, repertoire matching, and unshared song casts light on the benefit of a repertory comprising a number of song types (Beecher and Campbell 2005). This fundamental puzzle of birdsong complexity operates, in the case of song sparrows, to create an audible arena of negotiation in which distinctive positions between neighboring birds are defined. Sparrows' song types could be represented as a Venn diagram with an oval enclosing the repertory of each bird, showing overlapping and non-overlapping regions, and additional ovals could be added to indicate a bird's relation with more than one neighbor. But such a diagram would not capture the processual unfolding of countersinging bouts in the arena, which makes of the song repertories a fluid medium for ongoing management of aggression or dear-enemy neighborliness. The song signs deployed in countersinging are indexes, events in an episodic process that point through learning and memory toward other aspects of the process. At the level of the whole repertory, moreover, the typologies manifest an incipient hyperindexicality, in which song types take on different meanings according to their places in an array of categorically distinct indexes, matched, shared, and unshared — something more than a mere smoke-for-fire kind of indexicality. The indexes are related in the array by virtue of the details of their combinatorial structures.

Where in all this is the sparrow interpretant? The simple answer is: everywhere, dispersed across the niche-constructive processes that enable the sung semiosis and the organismal capacities on which they rely; but we can enumerate ingredients of the interpretant more precisely than this. The attentional focus of each bird enables the perceptual recursion by which particular features of sensory data are extracted, then linked to one another in representational relation. As we know, such recursion founds the aspectual nature of any sign, in which vehicle and object capture one another in a relation of part to part. Not every birdsong means a level of aggression or threat. The sparrows' songs do so by virtue of features of them that demand attentiveness and create the typology of matching, shared, and unshared songs. These features redouble the nuance of the semiosis because they are not perceived in the nature of each song alone but in the relations of songs in the typology — the essence of their hyperindexicality. The aboutness of an individual song is a function of multiple birds' repertories.

This points to another ingredient in a sparrow's interpretant process. It involves not merely the learning of songs, an aspect of all birdsong, but the shaping of a repertory across a full year in interaction with neighboring birds' repertories. We saw at the end of Part II that complex learning and the episodic memory on which it rests are fundamental to indexicality, and here the demands on both are compounded, as sparrows learn their songs from tutors but also reshape their repertories in ongoing semiotic negotiations with them and other neighbors. This extends the simpler learning involved in a smoke-for-fire index since memory in that instance can lead to learning in episodic conditions that are in principle little changing, even unchanging. In song sparrows' learning, instead, episodic memory contextualizes indexical meaning in ways that remain flexible with each novel bout of countersinging, and the song typology can continue to develop and shift through at least a third of the life of an individual bird. The indexes that result can alter their meanings as an old neighbor disappears, a new neighbor presents new song types, and new songs not involved in a birds' original tutelage are perceived as matches for old ones. The formation of a song sparrow's semiosis is thus the product not merely of recursive attention, advanced learning, and episodic memory but of all these in connection to a malleable experience of social niche construction.

Though the sparrows' semiosis is distinct in the details of its sung sociality, its malleability is in keeping with findings from other species. A study of brown-headed cowbirds (*Molothrus ater*), for example, concludes that their "development is facultative — that is, social or vocal development is neither a fixed nor a centralized program. It is assembled on-line from recurring interactions with others in the social environment." This context-dependency is apparent across multiple aspects of the cowbirds' lifeways that are often assumed to be fixed at least by early ontogeny, including "song potency's correlation with mating success, intrasexual competition, song learning, female choice, or even the idea of a species-typical mating system" (King, West, and White 2002, p. 180). The signs exchanged among groups of cowbirds arise from processes molded and remolded according to lived social interactions and habitat conditions — a flexibility, the researchers propose, that may have come to be heightened in this species because it is a brood parasite, laying its eggs in other birds' nests and leaving its young to be reared not by conspecifics but by as many as two hundred other kinds of birds. In any case, as with song sparrows, these processes involve the creation of a metarelational percept connecting things in the social niche in meaningful relation. They are interpretant processes.

Case 2

In a small portion of the world's songbirds, mainly species native to the tropics and Australia, two birds join in singing coordinated duets (for an overview, see Catchpole and Slater 2008, pp. 215–19). Usually these are pair-bonded birds, attached in long-term, monogamous relations and occupying their territory year-round. Duetting has been associated with various aspects of the shared lifeways of these mates, and it probably functions in many ways across the species involved and perhaps even several ways within single species (Hall 2004). The most convincing functions assigned to it involve announcing and guarding the commitment between the mates and joint defense of their territory or discouragement of potential intruders. The overlapping of these suggests once again the difficulty of isolating neatly packaged functions for animal behaviors, at least at general levels.

Mated pairs of Australian magpie-larks (*Grallina cyanoleuca*; Figure 18.2) sing duets typically lasting five or six seconds in which the two birds synchronize their own contributions of syllables. The syllables they sing, drawn from a modest species-specific repertory, are not marked as male or female — there is little gendered difference between repertories in this species (Hall 2006) — but in duetting each bird contributes syllables different from its partner, distinguishing the duets from their solo songs, which repeat a single syllable from the repertory. The contrasting syllables in the duets are arranged in one of two forms (Rek and Magrath 2017): Either the birds alternate syllable-by-syllable (antiphonal duetting) or one bird sings a few syllables and the other responds with a few of its own (sequential duetting). The duets form the auditory part of a performance that includes a visual aspect: wing movements alternating between the birds in a fashion coordinated with one another and with the singing. The precision of the synchrony of magpie-larks' performances increases with their growing experience together as a pair, and experiments using both auditory playback and robotic birds have suggested that heightened coordination along three axes — singing alone, movement alone, and singing and moving together — elicits

Figure 18.2. Male magpie-lark (*Grallina cyanoleuca*), drawn by Virge Kask after a photograph by Graeme Chapman.

heightened responses from birds witnessing the performance (Hall and Magrath 2007; Rek and Magrath 2016; Rek 2018). Duetting pairs perform from prominent posts in their fairly open, eucalyptus-rich habitats, seemingly to ensure the clear visibility as well as audibility of the performances and thus enhance their efficacy (Rek and Magrath 2017).

All this suggests semiosis operating on more than one level: between the mates themselves in the coordinating of their performance and between the pair of performing birds and other birds witnessing them. Whatever meanings are conveyed by magpie-larks' audiovisual duets, they depend on the pair-bonded birds' coordination being perceived and gauged by potential rivals for mates and territory. The duets convey, at least, a meaning of strength in numbers—precisely, in the number two.

Recent research has uncovered an extraordinary manipulation by magpie-larks of this meaning that deepens its semiotic possibilities (Rek and Magrath 2017). It is not unknown among duetting bird species for one bird to sing a solo song that imitates the features of a duet. Individual magpie-larks perform such "pseudo-duets" when their partners are absent, especially when one of the pair is nesting and the other foraging. A single bird sings contrasting syllables instead of the repeated single syllable usual in solo songs, arranging them either antiphonally or, more often, sequentially. Given the general meaning of true duets, the purpose of performing a pseudo-duet seems to be to deceive listeners, signifying a coordinated coalition even though one bird is absent. This interpretation is supported by a major difference between true and deceptive performances: pseudo-duets are performed from hidden posts, not out in the open like true duets, and they are not accompanied by the wing movements of true performances. The whole visual component of true duets is suppressed, in what seems to be a calculated attempt to make the deception effective by concealing the absence of ingredients that would expose it.

The interpretant process in birds witnessing duets, whether true or pseudo-, shows all the general features named in song sparrows' interpretants: the recursive focus on stimuli, parsing aspects of them; the resulting part-to-part representational relation between those aspects and aspects of a broader social situation; and the complex learning and episodic memory of situations and events within an elaborated social context. Magpie-larks perceive, in the context of their own learned song repertories, aspects of the duets they witness that differ from solo songs and signify the coalition of the bonded pair. Additionally, their perceptions are nuanced in degree according to the coordination of the performance. The sign that results is indexical, pointing both to the coalition in its social, niche-constructive context, and more specifically to the strength and durability of the bond.

The interpretant formation of the singer of a pseudo-duet is a still more complicated episodic affair. The bird has learned a repertory of three to six syllables from which it constructs just so many solo song types, each made up of one of those syllables repeated. It has learned also, with growing precision, to deploy these syllables in contrasting fashion in duets with its partner, an act that involves a distinct interpretant in which a relation is constructed between the duet interaction (with coordinated visual display) and a life situation, creating a meaning: unity of the pair in readiness to defend territory, or the like. In addition, the pseudo-duet involves another interpretant, one step back in the signifying chain, which intersects the intersection of the first interpretant with singing and lifeways. This creates a new sign in which the first sign now functions as a sign vehicle, while the sign vehicle and object of the first sign now function together as the object. In the true duet, dual singing means coalition strength; in the pseudo-duet, solo singing means duet + coalition strength. From this recursive interpretant process emerges the new index, which signifies *for witnesses* the same thing as the first sign (coalition strength) — this is the effectivity of the deception — but signifies *for the deceiver* something else involving a concealment of vulnerability and the partner's absence, and an evasion of threat. We can adjust Figure 3.2 to give an impression of this process:

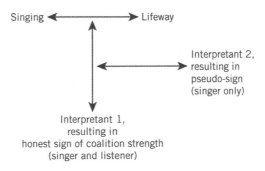

Figure 18.3. Recursive interpretant in the pseudo-duet.

Further, given the cognitive capacities on which semiosis is founded, we can fill out Figure 18.3 to indicate more than bare-bones semiotic

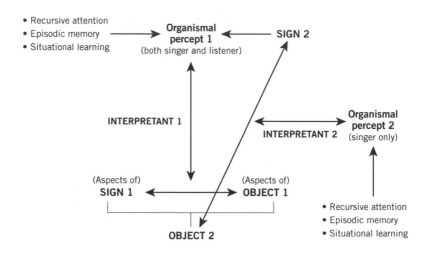

Figure 18.4. Redoubled sign structure, percepts, and cognitive foundations in magpie-lark pseudo-duetting.

recursion, showing the distinct percepts involved and general capacities as well as the compounded sign structures (see Figure 18.4). In principle, semiotic deception always requires recourse to such second-order interpretant formation on the part of the deceiver. The compounded situational or episodic learning involved in such signs separates them from different kinds of animal behavior, sometimes wrongly thought of as fully semiotic deception, that involve only mimicry of causal informational signals.

NINETEEN

Metarelations

in the Songbird Brain

How do the interpretants of song-signs form in a bird's brain? The song system explored in recent decades prompts some speculations. This system involves a number of clusters of neurons, or nuclei, in the cerebrum and thalamus of a songbird's brain, which are either absent or less developed in birds without vocal learning capacities (for summaries, see Jarvis 2004; Catchpole and Slater 2008; Mooney, Prather, and Roberts 2008; Moorman, Mello, and Bolhuis 2011). The system is, in other words, a distinctive neural architecture highly developed only among birds that produce song. These special *vocal pathways* connect with pathways for auditory reception and processing, also involving many nuclei — *auditory pathways* that share their basic architecture with non-vocal-learning birds and are not usually considered part of the song system proper. Auditory pathways are shown in Figure 19.1a, the special vocal pathways of the song system in Figure 19.1b.

In the song system (Figure 19.1b), axons of each nucleus project to one or more other nuclei, creating two primary pathways comprising eight nuclei. The pathways have several names, for example, the anterior and posterior vocal pathways, according to their positions in the brain (Jarvis 2004), and the learning and motor pathways, according to their general functions of song learning and song production (Catchpole and Slater 2008). I use commonly accepted names, the *song*

A. Auditory pathways in the songbird brain

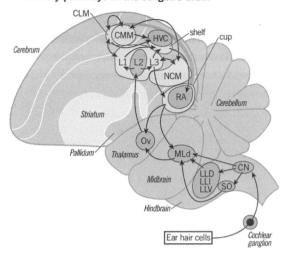

B. Vocal pathways in the songbird brain

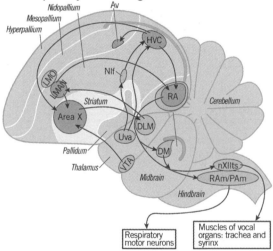

Figure 19.1. Connectivity in auditory and vocal pathways in the songbird brain (sagittal section, anterior to the left). Letters identify nuclei in the pathways, words different brain regions in which they occur. Major nuclei are identified in the text. A: Auditory pathways; major nuclei in light grey. B: Vocal pathways; major nuclei in dark grey. Drawn by Virge Kask after Robert C. Berwick, Gabriël J. L. Beckers, Kazuo Okanoya, and Johan J. Bolhuis, "A Bird's Eye View of Human Language Evolution," *Frontiers in Evolutionary Neuroscience* 4 (2012).

motor pathway and *anterior forebrain pathway* (Mooney, Prather, and Roberts 2008).

The *song motor pathway* is generally associated with the vocal production of song. Of its four main nuclei, the best understood are HVC (used as a proper name, though derived from an acronym for the brain region in which it occurs), which exerts a paramount control over song production, and the RA (robust nucleus of the arcopallium), which receives output from HVC and sends neuronal impulses through the midbrain, hindbrain, and spinal cord to muscles of the syrinx, trachea, and respiratory system.

The *anterior forebrain pathway* includes another four nuclei, three in the cerebrum and one in the thalamus, positioned beneath the cerebrum on the way to the mid- and hindbrain. This pathway is associated with song learning, plasticity, and variability, both during and after the juvenile sensitive phase, but its exact functioning has been harder to track than that of the song motor pathway, as the name of its most prominent nucleus suggests: Area X.

Both pathways of the song system receive sensory input from *auditory pathways* (Figure 19.1a), which process audible input. The main auditory pathway ascends from the ear through the hindbrain, midbrain, and thalamus to the pallial region of the cerebrum (related in cognitive function to the mammalian cortex; see Jarvis 2004), where it terminates in a region called Field L, similar in function to the mammalian primary auditory cortex. In songbirds, distinct regions in Field L (L1, L2, and L3 in Figure 19.1a) and adjoining it (CMM, caudal medial mesopallium, and NCM, caudal medial nidopallium) are associated with the learning and processing of songs from tutors and other conspecifics, and one of them has been proposed as the neural substrate of memorized songs (Moorman, Mello, and Bolhuis 2011). Experimental evidence supporting this points to a process of song memory formation involving immediate early genes expressed in response to heard song and leading to rapid neuronal changes in this region. Moorman and coauthors, moreover, liken the memory formation here to that in the rodent hippocampus, and we have seen that this region is strongly implicated in episodic memory (Section 14; also Allen and Fortin 2013). In general, processing of a heard song through Field L and adjoining regions seems likely to be involved in the construction of species-specific song percepts, with each area responding successively to specific features of the song (Jarvis 2004). As Figure 19.1a shows, regions adjoining Field L connect to the song motor pathway of Figure 19.1b, projecting directly and indirectly to HVC and RA (see Mooney, Prather, and Roberts 2008).

This is the barest overview of the intricate connectivity of the song system and the auditory pathways, and finer mapping resolution shows additional nuclei, connections, and loops that include more detailed networking between Field L and other nuclei (some of it indicated in Figure 19.1a). Both in summary and in detail, the map yields three important generalizations. First, the two song pathways, the anterior forebrain and song motor pathways, are not distinct pathways as was thought when they were first identified and given their names. Instead they are connected through circuits involving many nuclei and several brain areas. HVC projects not only to RA within the song motor pathway, but also to Area X in the anterior forebrain pathway. These projections in two different directions are mediated by interneurons, linking or interfacing neurons within HVC. Area X, meanwhile, projects to a nucleus in the thalamus (DLM, medial dorsolateral nucleus), which in turn projects to another nucleus of the anterior forebrain pathway, back in the cerebrum (LMAN, lateral magnocellular nucleus). LMAN sends impulses both to Area X, closing a loop within this pathway, and also to RA in the song motor pathway. The whole network shows direct output from HVC in the song motor pathway to the anterior forebrain pathway and indirect output through the thalamus back from that pathway to the RA, the nucleus that innervates the muscles needed for the motor control of song.

Second, though the nuclei of the song system are mostly found in pallial (cf. cortical) portions of the cerebrum (hyperpallium, mesopallium, and nidopallium in Figure 19.1b), they extend beyond it. HVC receives input from a nucleus in the thalamus (Uva, uvaformis), both directly and indirectly through another cerebral nucleus. Area X itself is located in the striatum, still in the cerebrum but related to a structure in the mammalian basal ganglia rather than the cortex, and its output, in the indirect loop that leads back to RA, is to DLM in the thalamus. Thus the pathways form a circuit from the pallial region of the cerebrum through the striatum of the cerebrum to the thalamus and back to the pallium. In mammals, a similar loop through distinct brain regions, from cortex to basal ganglia to thalamus and back to cortex, is implicated in fundamental capacities such as attentional focus, voluntary motor behavior, procedural memory, and learning. So the loop seems to be in the generation of birdsong also.

We encountered such thalamocortical loops in Section 13 and saw that they are also implicated in generating complex, situational attention. The thalamus is especially important, functioning as a mediating hub that differentially weighs sensory stimuli at the same time as it is differentially controlled by working memory. This interaction brings some information to salience in ways appropriate to the momentary, situational interests of an attentive animal. Given the situational attention in the processing and deployment of birdsong we witnessed in Cases 1 and 2 in Section 18, it is unsurprising but highly suggestive to find thalamocortical loops recruited in the song system.

What is most striking about the song system, finally, is its intricate, indirect, multiple connectivity. The loop within the anterior forebrain pathway, the loops back and forth between it and the song motor pathway, the pathways from auditory processing, learning, and memory areas in the cerebrum to the song motor pathway, and the pallial-striatal-thalamo-pallial circuits these networks traverse — all these reflect network parallelisms and series in birdsong processing and production. In general, looped series in neural networks across distinct nuclei or modules are opportunities for mediation of sensory input and feedback modulation of neuronal output, and they are basic to the models for the generation of situational attention and its mediation with episodic memory examined in Sections 13 and 14. These network complications point toward feedback mediation in the processing of memorized, learned song (input), with the potential to modulate the effects of HVC leading toward vocalization (output). The modulating feedback from the anterior forebrain pathway to the song motor pathway seems to enable plasticity and variety in song production, while the thalamus functions in focusing attention on external events and stimuli.

From all this, we can propose compelling congruencies between features of the song system, with its associated networks, and the foundational components required for interpretant formation. *Selective, situational attention* arises through thalamic participation and in loops between thalamus and pallium (cf. the mammalian cortex). *Advanced, episodic memory* is associated with brain areas analogous to the mammalian hippocampus, located close to the presumed seats of song memory around Field L. *Fine-tuned parsing of sensory input* into its individual aspects — key to the recursive, aspectual nature of signs — is created by thalamic selectivity or salience-filtering and the successive processing in subregions of Field L, in the pallium. *Learning shaped in depth* and detail by such aspectual, multi-dimensional percepts occurs in Field L and the anterior forebrain pathway. *Voluntary response* to new input in the context of such learning is produced in the song motor pathway, controlled especially by HVC. And *complex modulation* of this response across multiple mediations reflects the activity of the entire network and especially of the anterior forebrain pathway.

These congruencies suggest that we might discern in the networking of the song system the neural substrate for song-interpretant formation — always with the reminder that this does not take the form of a fixed architecture in the brain but is, like any brain substrate for experience, a set of processes set in motion across neuronal networks in interaction with extra-brain stimuli. The action of these processes is revealed in bird behavior and specifically, in Cases 1 and 2 in Section 18, in the operation of advanced situational awareness and episodic memory involved in deploying songs as signs in social contexts.

Can we dig deeper and see the metarelation basic to semiosis emerging, almost in a mechanistic fashion, in the serial and parallel interfaces of song processing? The metarelation seems to call for second-order processing in which distance from the immediate neuronal impact of information is gained, making possible its relation to other information. This suggests that the metarelational quality of the interpretant, and hence of the signs that result, depends on mediating loops and feedback circuits in the processes of cognition — folds, so to speak, in the processing circuitry. Such folds are indicated in the distance gained from an initial birdsong percept as it is circulated and recirculated through networks of attentional focus, episodic memory and learning, retrieval, and matching. The whole system enables the formation of and response to a song-percept whose aspects can be isolated in cognition and matched to other aspects of the social niche. The percept, in itself nonsemiotic, enters into semiosis through this relation to a relation formed in cognition.

The metarelation, then, can be seen as emerging from a weighing of already processed neuronal impulses in additional systems of neurons that feed back to the source of the first impulses. The song is gauged, its impact modulated and measured against other stimuli brought to salience by focused attention. In this process, plasticity of output (anterior forebrain pathway) is balanced against learned, remembered fixities (Field L) and crystallized motoric complexes (song motor pathway) — all to enable niche- and situation-appropriate response. The interpretant comes into being, and with it semiosis in all its rich potential.

Case 3

Our third case study looks inside the female songbird brain to a set of responses to particular kinds of male song. While it cannot confirm a model of the cellular mechanisms of interpretant formation, it indicates that semiotic folds and metarelational mediation may be recruited even where simpler signaling seems to be at work.

In the 1990s it was found that female domesticated canaries (*Serinus canaria*; Figure 19.2) respond to particular syllables in some male song with an increase in "copulation solicitation displays" (Vallet and Kreutzer 1995, Vallet, Beme, and Kreutzer 1998; for a review of earlier research, see Leboucher et al. 2012). The syllables in question, referred to as "sexy syllables," are phonologically distinct from other syllables in canary repertories, involving fluctuation between two notes of widely divergent frequency (hence wide "bandwidth") produced by different sides of the syrinx, unlike one-sided, unsexy syllables. The fluctuation, typically at a rate of more than fifteen times per second, requires precise coordination of respiratory and syringeal muscles, rapidly opening and closing each side of the syrinx and alternating pulsating expiration with "minibreaths" gulped between each pair of notes (Suthers, Vallet, and Kreutzer 2012).

Figure 19.2. Female canary (*Serinus canaria*), drawn by Virge Kask after a photograph by Yeray Seminario.

In playback experiments, heightened female response has been correlated with several individual parameters of sexy-syllables: wide bandwidth or distance between the two notes of the syllable, overall low frequencies or pitch range, duration or temporal length of each syllable, and amplitude or loudness. Though learned female responses to male song are amply attested for canaries (Leboucher et al. 2012), these preferences involving sexy syllables are in part predispositions evoked regardless of prior learning or social or sexual experience (Pasteau et al. 2009; Suthers, Vallet, and Kreutzer 2012). This is in keeping with a number of other species, where female preferences have been generally correlated with longer songs, increased overall song output, and higher complexity of individual songs or of the complete repertory (Catchpole and Slater 2008; Pasteau et al. 2009).

Expansion of all these parameters — length of song, size of repertory, and complexity of songs or syllables — presumably demands extra expenditure of energy and resources on the part of male singers, and this has led researchers to assign them to the category of "honest signals" operating in the dynamics of sexual selection. Such signals, hard to counterfeit because of the difficulties involved in producing them, form reliable, stereotyped indicators to their receivers and are widespread elements of animal communication. Like other kinds of signals, they often involve thresholds at which preprogrammed responses are set in motion, and, though they can incorporate signs and interpretant formation, they need not — and mostly, across the animal kingdom, do not. In the honest signal hypothesis, male canaries' sexy syllables, like other song complexities, enter into sexual selection as indicators to female auditors of the strength, vigor, or quality of a potential mate (Catchpole and Slater 2008; Suthers, Vallet, and Kreutzer 2012), analogous to the ability of a peacock to lug around spectacular tail plumage.

The story of these syllables, however, has turned out to be more than a narrative of honest signaling and largely innate response. A copulation solicitation display in response to song, like any complex behavior, is the tip of an iceberg, with a cascade of neuronal processes behind it reaching, in the case of sexy syllables, to the highest levels of the song system. Female canary responses, which researchers measured at first by the frequency of their display, soon came to be gauged through bioassays of several sorts, with revealing results. Researchers discovered that females showing more discrimination between sexy and unsexy songs and performing more frequent displays tended also to have larger HVCs in their own song motor pathways, suggesting an unanticipated connection of this production pathway to song perception (Leitner and Catchpole 2002). The analogous correlation in mature male canaries between the proportion of sexy syllables in their songs and the sizes of their HVCs was more easily explained by the song production role of that nucleus (Leitner and Catchpole 2004).

Next, an immediate early gene called ZENK was found to be induced in females hearing sexy syllables, at levels higher than in those hearing songs without them. ZENK is important in many animals in the downstream regulation of other genes, with effects that have been associated with neuronal plasticity and the memory formation and learning dependent on it (Knapska and Kaczmarek 2004). Increased induction of it is known to be associated with song hearing in several songbird species, including both female and male zebra finches exposed to courtship songs (Avey, Phillmore, and MacDougall-Shackleton 2005; Mello, Vicario, and Clayton 1992; and McMillan et al. 2017). In canaries, the increased ZENK induction is especially marked in the CMM, a nucleus we encountered before at the pallial apex of the auditory pathway that projects to HVC in the song motor pathway (Leitner et al. 2005; see also Haakenson, Madison, and Ball 2019; Figure 19.1a). This points to a link between sexy-syllable responses and the juncture of the song motor pathway and the auditory pathway, with its learning and memory functions and its contribution to the species-specific recognition of song.

Hormone levels too are affected in female canaries in response to sexy syllables, specifically, levels of neurosteroids in their brains.

Heightened testosterone levels were first detected in the canaries' bloodstream and yolks of eggs they laid (Catchpole and Slater 2008; Leboucher et al. 2012), and more recent work has found rises in the estrogen estradiol as well (Haakenson, Madison, and Ball 2019). Testosterone is a metabolic precursor of estradiol, and in this response to song it might well function in a chain leading to the rapid synthesis of that estrogen, driving in turn changes in gene expression and then in cognition and behavior. Recent research — not on canaries but on female zebra finches' responses to male song — suggests exactly such a mechanism, localizing the production of both testosterone and estradiol in yet another nucleus in the auditory cortex near the juncture with the song motor pathway, the NCM (Bournonville, McGrath, and Remage-Healey 2020; Figure 19.1a). All these hormonal, immediate early gene, and neurobiological results suggest a more active role for female canaries in processing sexy syllables than was at first suspected, a role involving their own song networks.

Since the 1990s, the experiments that gauge female responses to sexy syllables have typically used synthesized songs for their playbacks, in which a brief unsexy intro leads to the sexy syllables, followed by an outro of more unsexy syllables. Though all the individual syllables are derived from real canary song, they are computer-constructed into a pseudo-song unlike songs heard in nature. One of the most striking recent studies has departed from this model (Haakenson, Madison, and Ball 2019). These researchers first repeated earlier experiments with pseudo-songs, confirming the increase in ZENK induction in the CMM in response to sexy syllables presented in this unnatural way. Then they exposed two groups of birds to full, natural songs for two weeks, one group hearing songs with sexy syllables, the other songs with the sexy syllables removed. On the fifteenth day, each group was subdivided, with some hearing natural sexy songs, some hearing natural unsexy ones, and some controls hearing no songs. All the birds that heard song (as compared to those treated to silence) showed an increased induction of ZENK, correlated between the CMM and NCM, but there was no significant difference in ZENK production between the groups that heard sexy and unsexy full songs. The researchers speculate that the pseudo-songs, emphasizing sexy syllables in an artificial acoustic context, form a supernormal stimulus (Tinbergen 1951) that elicits a response exaggerated beyond those found in nature, suggesting that pseudo-songs are unreliable as a gauge of natural song communication. The richer acoustic context of the full songs, instead, may include signifi-cant, undetected features other than sexy syllables, engaging birds in a more holistic way and suggesting so far undetected complexities of song features involved in birds' percepts.

One more recent study complicates matters further with ques-tions of the female canaries' own vocalizations (Amy et al. 2015). These come in several types, ranging from simple calls and trills — not full birdsong at all, but the kind of vocalizations that virtually all birds produce — to "female-specific trills," soft, complex vocalizations consisting of more than one note and associated with nonvocal mat-ing routines. The researchers found these special vocalizations to be correlated with copulation solicitation displays in female responses to sexy syllables, but not in a linear association. Instead, the perfor-mance of each, the displays and the trills, increased in frequency independently with the "attractiveness" of the male song heard, mea-sured according to the rapidity of sexy-syllable performance in it. Each showed, moreover, a falling off in frequency with successive repetitions of the male song, a habituation effect that was however less marked for songs of higher quality, thus seemingly reflecting the perception of particular song features. None of the other, sim-pler kinds of calls and trills showed any of these correlations, either with the sexiness of the male song or the frequency of copulation solicitation displays. Finally, the female-specific trills, alone among female canaries' vocalizations, are sung *during* the male song, begin-ning soon after its start, in a kind of irregular, unsynchronized duet. Simple calls, when they are heard in proximity with the male song at all, invariably come after it.

We can tease several general messages of widening import out of all this work. First, female canaries are up to behaviors considerably more flexible and involute than a stereotyped response to an honest signal. The involvement in female response of the song motor pathway (specifically, HVC) and the auditory pathway at its meeting with the song motor pathway (CMM, NCM) suggests that learning and memorization play a role in modulating the females' innate perceptual predispositions, as they do in other ways in all birdsong. Females' vocal exchanges with male songs, virtually ignored in the literature until recently, suggest one explanation for the enlarged HVCs of more responsive individuals: HVC might function not in a purely perceptual role but in a role associated with song production, as in males. Females' gauging of the quality of male song elicits not only visual displays but also vocalizations that interact directly with those of potential mates, responding to males' signs with signs of their own, and eliciting from males interpretant formations to match to their own as the precursor to counterresponses.

The complexities of female canaries' responses to male song push us to recognize unanticipated subtleties of perception at work in songbird sociality. Whether or not the females are dependent for their responses in the wild on subtle differences in sexy syllables, it is clear that they can register these differences at a level far more sensitive than human auditory perception allows. This supersensitive perception of individual syllables has recently been detected also in zebra finches, which turn out to be more sensitive to small alterations in the phonology of individual syllables in their songs than to the sequence in which the syllables are presented (Fishbein et al. 2019). This does not invalidate the importance of sequencing in all birdsong, and the case of duetting magpie-larks (Case 2) provides an instance where sequence is central; neither does it push the finches' syllables onto the semantic terrain of words in human language — it does not make their song compositional (see Section 17) — because there is no compelling evidence for individual meanings attached to syllables. Instead, it suggests an additional level of structural differentiation in birdsong from which interpretants are formed and indexes come to point to their objects. Differences between the sequential order of syllables now stand alongside more subtle differences in the articulation of particular, learned syllables, enriching the hyperindexical communication of birdsong and boosting it onto a perceptual level that is literally superhuman.

Finally, the sexy-syllable case shows the immediate impact of sung sociality on gene expression in canaries' brains. Female responses involve the learned and remembered concomitants of the song system and the gauging of sensory input in its circuitry. At the same time, they also entail neurohormonal regulatory pathways and altered genetic expression, under the impact of input from the birds' rich social niches. This affirms the connected network reaching from the niche to genes but in a different sense from classic niche-constructive theory, where it was thought to operate across long *durées* of selective timescales. The radical version of niche construction raised by the third abstract machine of evolution (Sections 6–7) instead shows the connection bringing about momentary shifts in genetic expression and resulting in momentary modulations of behavior. From social input (male song) to social output (female responses in all their variety) there extends a complexly mediated dispositive that engages attention, learning, episodic memory, and reciprocal behavior in ways that involve not only epigenetic but genetic plasticity. All this, in the case of songbirds, is set in motion as part of a semiotic process.

Nonsignifying Marvels:

Honeybees

TWENTY

Puzzles of Invertebrate

Complexity

Insects, Darwin knew, posed perilous test cases for natural selection. The *eusocial* structure of ant colonies, which gives rise to castes of organisms incapable of reproducing, flew in the face of inheritance, a basic component of his abstract machine. The combs that honeybees construct, mathematically ideal for maximizing storage space while minimizing wax demands, seemed to many in his time inconceivable except in the light of divine design. Darwin pondered how these phenomena might have sprung "from a few very simple instincts" through accumulated, incremental changes, and his reasoning makes for incandescent reading in *The Origin of Species* (Darwin 2003, chap. 7).

Complexities of insect behavior and social organization have continued to tax evolutionary thought since Darwin, and his focus on eusocial insects has been followed so often that ants and bees — two groups in the order Hymenoptera, a family in the case of ants (Formicidae) and a superfamily (Apoidea) comprising seven families in the case of bees — rank among the most thoroughly studied of all invertebrates, indeed all animals. Their colonies are now called *superorganisms*, since the divisions of labor in them involve distinctions of function among whole organisms analogous to those among organs in an organism (Wilson and Sober 1989; Hölldobler and Wilson 2008). This higher-order functionalism, however, particularly its division of reproductive labor, requires selection acting at the level of the group rather than the individual organism. Darwin saw this as he considered nonreproducing castes of ants, mooting the idea of "selection . . . applied to the family as well as to the individual" (2003, p. 150). But the gene-centered thought of the mid-twentieth century struggled to offer a mechanism for group selection, and much ink was spilled from the 1960s on over the resulting levels-of-selection debate. Today, extended evolutionary thinking adopts an inclusive, processual view in which levels of selection are understood to shift now and then along with major evolutionary changes (Okasha 2006). The transition that introduced multicellular organisms, for example, required the suppression of competing selection at the cellular level (Maynard Smith and Szathmáry 1995), expanding the borders of what could be considered a functionally delimited organism to include not only single cells but clusters of differentiated cells. The processes at work between a superorganism and its constituent organisms bring about a similar shift.

Alongside evolutionists, many twentieth-century ethologists turned their attention to ants and bees, most famously Karl von Frisch, awarded a Nobel Prize for his decipherment of the "waggle dances" of honeybees, and E. O. Wilson, who took from his studies of ant societies the inspiration for a whole new field, sociobiology (von Frisch 1967; Wilson 1975). Their analyses of bees' and ants' social complexity decisively refuted earlier views of insects as "simple and small reflex automata" (in the characterization of Menzel and Giurfa 2001, p. 62) or "reflex machines" (Leadbeater and Chittka 2007, p. R703). Often, however, it veered in the opposite direction, fostering anthropomorphic tendencies — a drift toward humanist parochialism that moved far, proposing insect capacities such as evaluation, consideration, decision making, concept formation, language, intentionality, and symbolism. These tendencies are still alive today (for example, see Menzel 2019). Like other versions of humanist parochialism, however, this one can hamper the understanding of the nonhuman difference it aims to explain. To resist it, I'll follow recent trends in insect research that exploit new technical possibilities in investigating the cellular, neuronal, biochemical, and genetic foundations of insect sociality, tracking its pathways out from DNA molecules and neural substrates to the constructed niche, and in again from niche to DNA.

We have followed some analogous pathways for songbirds, and comparison of those with insect pathways will reveal a basic difference. Birds and insects sit on opposite sides of the divide between semiotic information — the realm of meaning and aboutness — and information as causal covariance. On top of bioinformational channels extending across all life-forms, birds overlay cognitive processes that create signs and meaning. Nothing like these metarelational dynamics goes on in insects' brains, for all their stunning operations. Indeed the wonder of insect behaviors is that they exploit causal bioinformational systems to construct immense social complexity *without any meaning at all*. Darwin already intuited this in his hypotheses about honeycombs and castes, and recent research has put his intuition on a firm footing, connecting the microcosm of insect genes and neurons with the macrocosm of their ecological and social niches in new ways and with new, granular details, but without any sign of semiotic cognition.

Among much-studied insects, the honeybee species *Apis mellifera*, originally found in Europe, Africa, and southwest Asia and now dispersed worldwide in about twenty subspecies, is probably the most studied, exceeding ants and even the fruit fly *Drosophila melanogaster*, widely used for genetic research since Mendelian experiments on it began a century ago. There are obvious reasons for such research interest since honeybees are across the world the chief pollinators for a large portion of human foodstuffs. On a less pragmatic level, fascination attaches to the superorganismal colonies they form, which challenge us always to adopt a bifocal view, divided between individual bees and whole colonies. The middle ground here forms the operational heart of the honeybee dispositive, structuring the colony and determining the roles of bees in it. Looking outward from it, its mediated systems of feedback mold honeybees' niche construction as a whole, extending to their coevolution with flowers and beyond it to its broader ecological consequences. Looking inward, the systems shape individual bees down to the molecular level, activating evolved modes of neuronal and physiological plasticity. These involve pheromones and related chemicals, immediate early genes, and the altered expression they bring about of other, "late" genes.

Without taking account of this middle ground we cannot explain the most basic and interesting aspects of honeybees. These include control of individual organismal reproduction; dynamics of swarming and hive division — *super*organismal reproduction; feeding and care of the brood through larval and pupal stages by female workers; transformation of workers across several stages devoted to different tasks, ending usually in a foraging stage; foragers' provisioning of the hive with pollen, nectar, water, and plant resins; and homeostatic regulation of foraging levels in response to in-hive and outside conditions for efficient resource exploitation. All these take shape in the social interplay of individual and colony in the niche, generated through the abstract machine of hypermediated regulation. They are processes emergent from the dispositive as a whole.

There are several surveys of these various aspects of honeybees and their colonies. One of the best is Thomas D. Seeley's *Honeybee Ecology: A Study of Adaptation in Social Life* (1985), which he followed a decade later with a detailed analysis of food-collection mechanisms in the superorganism, *The Wisdom of the Hive: The Social Physiology of Honey Bee Colonies* (1995). These are noteworthy not only for their authoritative coverage, ingenious experimental designs, and acute analyses of experimental results but also for the shift of emphasis they reveal across the decade separating them. *The Wisdom of the Hive* moved farther than *Honeybee Ecology* toward an understanding couched in terms of information transmission and the feedback economies it generates. Seeley showed that the solutions to the problem of provisioning the hive, a challenge involving thousands of foragers and shifting needs for and availability of resources, result from distributed responses to simple cues, with minimal information transmitted from bee to bee — a decisively nonhuman form of wisdom.

Research over the last twenty-five years has confirmed and extended this view of the honeybee superorganism and its implications for the capacities of individual bees. In the following sections I survey these findings along three axes: the biochemical controls mediating honeybee sociality, the waggle dance and forager allocation across food sources, and the nature of individual bee cognition.

TWENTY-ONE

Sociochemistry

of the Superorganism

In the light of the complexities of regulatory mediation between individual bees and the colony — the third abstract machine of evolution at work, connecting genetic molecules to niches through animal sociality — the idea of *sociochemistry*, a chemistry of social interactions (Slessor, Winston, and Le Conte 2005), is loosed from the taint of reductive biochemical determinism; likewise is *sociogenomics* distanced from genetic determinism, the founding weakness of the sociobiology from which it took its start (Robinson, Grozinger, and Whitfield 2005). Biochemicals and genes are now understood to be elements regulated through feedback circuits even as they regulate; they are controlled even while they control. There is no originary locus of action or control, hence no determinism, only loops of interaction spanning the dispositive, from the niche to DNA. Indeed there never was such a locus in life's history, only circuits redoubling their intricacy across 3.5 billion years of evolution.

Today we understand the reciprocal processes at work in honeybees as well as in any other social organism. In their most general outlines these are features of all organisms. The genome produces the full repertory of proteins in an organism, its *proteome*, but the proteome at the same time controls the genome and regulates gene expression through multiple layers of protein functions. These involve, at the DNA end, transcription factors enabling or inhibiting the transcription of nucleotide sequences onto messenger RNA. *Transcriptomics* is the new science analyzing the changing repertory of mRNA (the *transcriptome*) that results from this shifting operation. Additionally, working with proteins within an organism as partners in their gene regulation are various kinds of circulating signal chemicals gathered under the catch-all term *hormone*. In the higher-order functionalism of insect superorganisms, similar chemicals excreted and passed between individual insects rise to special prominence, functioning like interorganismal hormones. These *pheromones* provide molecular signals in the form of stimuli external to the individual organism.

In a honeybee colony, pheromones are dispersed, by physical contact among bees and as volatiles in the air, through the hive and all the way to flowers visited by bees foraging for nectar and pollen. Their production occurs in the endocrine systems of individual bees and is governed by genetic instructions, but proteins regulate the instructions and are in turn regulated by the circulating pheromones themselves. Pheromone molecules encountered by individual bees bind with receptors on cell membranes and set in motion cascades of internal cellular responses along signal transduction pathways, often ending in protein transcription factors that alter gene expression. But this is no true ending, since the genetic alterations have reciprocal effects out through cell and organism, changing behavior and pheromone transmission. Sometimes these effects take the form of slow developmental and physiological changes — pheromones involved in these are called "primer" pheromones — and sometimes they bring about rapid behavioral changes — "releaser" pheromones. Some pheromones show both primer and releaser effects in different contexts. In the releaser pheromones the threshold action typical of biochemical signals is especially clear, but it is involved also in primer pheromones, and the pathways leading to these two different kinds of effects can be similar at molecular levels (Bortolotti and Costa 2014).

By virtue of their dual impact, within the cell and between individual bees, pheromones constitute links in bioinformational pathways from genes to cell, cell to organism, and organism to superorganism. By virtue of their regulation by other factors in these loops, pheromones are responsive to changing external conditions, forming a bridge between niche dynamics, including social interactions, and the loops of the dispositive. These pathways have been envisaged as vectors running in opposite directions, from social behavior to genes and genes to society (Robinson, Fernald, and Clayton 2008), but this does not firmly enough counter the old, unhelpful dichotomy of nature and nurture. It is more accurate to conceptualize them as aspects of a regulatory network encompassing genes, gene products including gene regulators, signal chemicals internal and external to the organism, organism, superorganism, and niche (Linksvayer et al. 2012; Laubichler and Renn 2015). There is no possibility in this network of direct determination of social behavior by genes. There are only mediated effects, conducted and transduced along the networked paths. Likewise, there is no possibility of direct social impact on genes, only more mediation. Between these distant points, the network *always mediates* through processes molded by the basic abstract machines of evolution.

In the space between organism and superorganism, the honeybee dispositive depends on the differentiation of several castes of bees as well as on the physiological plasticity present in all of them but especially marked in one, the workers. The castes include the queen, normally the only reproducer in the hive; male drones that mate with queens from other colonies; and female workers variously tasked with cleaning the cells in the honeycomb, feeding and caring for the brood of eggs, larvae, and pupae growing in the comb, constructing, maintaining, and guarding the hive, and foraging, especially for nectar, but also for pollen, water, and tree resin. These different worker functions emerge across a worker's life roughly in the order given, though there is much variety in the timing of their appearance — or their appearance at all — in individual workers (Seeley 1995, pp. 29-31). The workers sometimes lay unfertilized, haploid eggs (having a single set of chromosomes), which become haploid male drones. Their only function in the superorganism is to mate. The female workers themselves are instead born of the fertilized, diploid eggs of the queen.

Four broad economies intersect in this dispositive: reproduction *in* the hive, whereby workers (from the queen) and drones (from the workers) are generated; reproduction *of* the hive, whereby a new superorganism is created by the fission of an older one in a process involving swarming; provisioning of the hive by foraging workers; and maintenance and protection of the hive. Pheromones form essential links in the networks holding all four economies in homeostatic balance with each other and with the changing niche. They include more than fifty signal chemicals differentially produced by the queen, the larvae, and the workers and joined into a number of complex cocktails. (Drones generate few pheromones, most of them involved in sex, and play only an indirect part in these homeostatic systems.) Our understanding of the operation of these chemicals has burgeoned in the last few decades (for reviews, see Slessor, Winston, and Le Conte 2005; Alaux, Maisonnasse, and Le Conte 2010; Bortolotti and Costa 2014).

Some pheromone effects seem relatively simple, though the cascades involved at the molecular level are not. Alarm pheromone, a part of the economy of hive protection and maintenance, consists of about fifteen compounds released by a young worker's sting gland that attract other workers to attack the intruder she has stung. The individual roles of its component compounds in the molecular changes fostering this behavior are ill understood, but we know enough to know they are intricate and multiple (Slessor, Winston, and Le Conte 2005; Bortolotti and Costa 2014). Even a brief exposure to one of its main compounds alters transcriptional factors in the antennal lobes, the olfactory centers of the bee brain, activating fight-or-flight and aggression pathways in other, higher brain regions for a quick response to the threat. An hour after exposure, the resulting cascades have up- or downregulated the expression of hundreds of genes, laying a foundation for heightened interorganismal responsiveness to subsequent alarms that has been compared to the intraorganismal action of T-cells in the human immune system (Alaux, Maisonnasse, and Le Conte 2010).

Reproduction in the hive is normally the province of the queen alone, which she controls with her own special pheromone cocktail, called queen mandibular pheromone (QMP) because the gland that produces many of its components is located in her mandibles. This "primary superorganismic pheromone" (Alaux, Maisonnasse, and Le Conte 2010) is composed of more than a dozen compounds, and its functioning is correspondingly varied. It attracts a retinue of workers to the queen. They lick and antennate her and then disperse her pheromone through the hive, signaling to the colony the presence of an egg-laying queen. QMP enforces her reproductive monopoly through changes it brings about in the regulation of hundreds of genes in the workers (Slessor, Winston, and Le Conte 2005; Alaux, Maisonnasse, and Le Conte 2010). These inhibit the development of workers' ovaries, their construction of the special honeycomb cells needed for breeding new queens, and their production of the "royal jelly" that induces larvae to develop into queens when it is continuously fed to them. The exact makeup of the QMP cocktail, also, shifts with the aging of the queen, who may live for several years, unlike workers, who typically live for several weeks. She mates only at the beginning of her life, with many drones, storing their sperm and later laying hundreds of eggs a day. The changes in QMP makeup seem to correspond to these life-cycle changes, at first mainly attracting drones (Bottolotti and Costa 2014) and later controlling the reproductive physiology and behavior of the workers. QMP also plays other roles in other social contexts, as we will see.

The larvae join synergistically in strengthening the queen's reproductive monopoly, producing their own pheromone, brood pheromone (BP), a mixture of ten compounds with multiple effects, especially on the workers in closest contact with the brood (see Pankiw 2004). In high doses BP fosters a large number of nurse workers, increasing the area of the comb devoted to brood rearing, and like QMP it inhibits the development of worker ovaries (Bortolotti and

Costa 2014), functioning in this along with a separate larval phero-mone, E-β-ocimene.

The effect of these pheromones on workers is an especially complex one, since they also help to regulate worker maturation. Workers' age-related changeover from in-hive or near-hive func-tions — maintaining combs, nursing, regulating hive temperature, guarding the hive entrance and more — to foraging is not automatic, and its timing and demographic balance are crucial to the homeo-stasis of the superorganism: more foragers means more resources for the colony but fewer bees devoted to in-hive tasks, and vice versa. In nurturing the brood this is an especially delicate balance, since larger broods require both more pollen to consume, gathered by foragers, and also more nurses to feed them. The action of BP is cor-respondingly varied: low doses of BP decrease the age of the foraging transformation, resulting in more foragers, while high ones increase it, leading to more nurses. E-β-ocimene, meanwhile, promotes the transformation to foraging. The secretion of both pheromones shifts also according to the age of the larvae, regulating the foraging/nurs-ing homeostasis in this way too: young larvae, with little need for nursing, produce little BP and more E-β-ocimene, promoting forag-ing, while older larvae produce more BP, promoting nursing (Borto-lotti and Costa 2014).

Here we have crossed over from one economy to another, from in-hive reproduction and brood nurturing to the provisioning of the hive by foragers, and both queen and brood play important roles in modulating the balance between the two. QMP and BP decrease the levels of juvenile hormone in workers, a hormone important in their maturation and transition from in-hive labor to foraging. They seem also to alter the workers' response threshold to sucrose which, when low, conduces to foraging (Alaux, Maisonnasse, and Le Conte 2010; for additional effects of BP on foraging, see Pankiw 2004). But the workers themselves, producing their own pheromones, prob-ably play the most important part in regulating the balance of nurses and foragers, and here a Malthusian feedback loop is clear. Foragers generate the pheromone ethyl oleate (EO), produced from fermented nectar in their honey crops, in higher levels than other workers, the queen, or the larvae. Returning foragers circulate EO especially through physical contact with the in-hive workers, and it stimulates the growth of a worker gland used in brood feeding and increases the average age of transformation to foraging (Alaux, Maisonnasse, and Le Conte 2010; Bortolotti and Costa 2014). Thus an abundance of returning foragers (abundance of resources, including pollen for the brood) results in an abundance of nurses (more feeders for the brood), until the foraging troop is depleted. Then the EO circulation decreases, and with it the age of transformation, so foraging numbers increase again. We will see later that this is not the only instance of homeostatic balances signaled by contact between returning foragers and workers in the hive: even the waggle dance loses its semiotic luster in the light of this mechanism.

In the economy of superorganism reproduction, finally, QMP again plays a central role. Its absence, through loss of the queen, induces workers to extend the feeding of royal jelly to certain larvae, nurturing new queens, and can result in development of some workers' ovaries (Slessor, Winston, and Le Conte 2005). Its attenuation, when the hive grows too large for its effective circulation, leads workers least exposed to it to leave the nest and swarm nearby — the first stage in establishing a new colony (Alaux, Maisonnasse, and Le Conte 2010; Bortolotti and Costa 2014). The old queen joins this swarm, and in the new nest her QMP will upregulate wax production and the workers' building of the comb (Alaux, Maisonnasse, and Le Conte 2010). Meanwhile in the old colony, deprived of the queen and her QMP, a new queen is reared. Swarming is another multipheromone phenomenon showing loops intersecting other loops, since the workers themselves produce in glands on the backs of their abdomens chemicals that induce other workers to join them (Slessor, Winston, and Le Conte 2005). We will return to the process by which the site of the new hive is determined when we take up the waggle dance.

This is a sketch only of the knowledge achieved in recent years of sociochemistry in the superorganism, and a fuller picture would include additional pheromones, more glands producing them, more mechanisms of their dispersion, and additional homeostatic balances they regulate (Bortolotti and Costa 2014). Even this sketch, however, conveys the challenges involved in trying to understand bees' pheromone communication. The fifty-plus substances active in it can seem redundant in their function, but it is more likely that nuances differentiating their multiple operations continue to elude detection, that these operations involve many distinct molecular pathways from sense organs to brain, still little understood, and that they form emergent systems with "dynamic properties requiring a complex systems approach for their analysis" (Slessor, Winston, and Le Conte 2005, p. 2740; see also Pankiw 2004).

The complexity generated along these pathways has often fostered the illusion of planning, social decision making, and conceptualization in the superorganism, but the chemical systems of honeybee homeostasis are at best metaphorically described and at worst misrepresented using such terms. To be useful here, even the term *communication* needs to be shorn of the associations it usually carries. The molecular pathways of honeybee sociochemistry invoke this concept only as we might use it to speak of two rooms in a house communicating with each other or two tunnels in a subway grid—not two intelligences. Sociochemistry explains honeybees' intricate social organization as the nondeterministic, reciprocal regulation generated in a network of chemical links. The network connects pheromones and behavior through mutual mediations communicating (in this restricted sense) among individual bees and beyond them with the superorganism and niche. Such explanation calls for no higher cognition, any more than higher cognition is required to explain insulin or testosterone regulation in my body—hormones in the organism, like pheromones in the superorganism.

Neither is there a place in this sociochemistry for the cognitive concomitants of semiosis and meaning. It is through and through a matter of signals, not signs. No attentional focus is involved—none of the recursive parsing of the world it enables — and no advanced, epi-sodic memory or learning capacities. Indeed, the forms of sociality structuring the superorganism require no awareness of perception at all, none of the folding over on itself by which brain circuitry brings about the sense of having a sensation. Where none of these aspects of cognition come into play, there is no interpretant formation or pro-cessing detour through perceived relations to relations — the metare-lational surplus fundamental to semiosis. What we witness instead is the nigh-miraculous organizing potential of hypermediated causal information. The third abstract machine of evolution is hard at work; of the fourth there is no trace.

TWENTY-TWO

Desemanticizing the

Waggle Dance

This reasoning, militating against meaning and semiotic capacities in the most advanced of social insects, does not take the form of an argument from parsimony of the sort often made by evolutionists. Occam's razor arguments pose questions such as this: If bioinformational pathways of pheromones, transcription factors, and immediate early genes can explain the homeostatic economies of the superorganism, what need is there to invoke a meaning-generating cognition regulating them? Or: What could account for the evolutionary emergence of such cognition, when the elaboration of signal pathways ubiquitous in the animal world maps a more direct road? Argument from parsimony, however, sits uneasily among the historical, provisional systems of biology; applied to them, parsimony seems to take on the status of a vital principle at work. In the case of a honeybee colony, where is the presumed locus of its action? It cannot be the superorganism alone, for there is no superorganism *alone*: It is a continuous process arising only in relations to its niche and to the individual organisms it comprises. Every other locus for parsimonious action we might name, from the bee down to DNA strands, is similarly imbricated. Perhaps, then, parsimony is supposed to act on the historical process by which all these things come into mutual dependence, but it's hard to conceive a principled reason why the consequences of such a process should be thus constrained. Occam's razor is best applied to human logic or, in the natural world, to phenomena congenial to reductive science, not to life-forms and their histories.

Instead of parsimony, the exclusion of semiosis from superorganism regulation relies on the construction of a model of this regulation and the evidence that supports it. Chemical signaling mechanisms form the central evidence, ever more amply yielding their secrets and explaining behavioral patterns in the colony. These patterns seem to include no role for higher cognitive controls, not because of a principle but as the result of an evolutionary history; and they leave little place even for proprioception on the part of individual bees.

To press the argument further we will take on, in this and the following section, the two foremost challenges to a nonsemiotic, nonmeaningful view of honeybee lifeways. The first of these is a specific and astonishing communicative behavior among honeybees, the *waggle dance*. Elucidated starting in the 1920s in the work of Karl von Frisch and his associates, this has gained pride of place among the wonders of animal behavior, marking an apex of invertebrate achievement (von Frisch 1967). Its communicative efficacy has rendered commonplace talk of meaning and introduced to honeybee research such terms as sign, symbol, syntax, vocabulary, and language; von Frisch from the first called it the bees' *Tanzsprache*. We will see, however, that the phenomenon operates outside the semiotic realm, as causal information without content. How can it do so?

The second challenge is more general and concerns the insect brain and the discoveries about honeybee cognition, memory, and learning that have emerged since the "cognitive revolution" in insect studies of the 1990s (Avarguès-Weber, Mota, and Giurfa 2012). How do these impressive capacities differ from those in birds that travel under the same names?

In the waggle dance, according to the now standard account that emerged from von Frisch's work, a foraging worker bee communicates to her nestmates information about the flowers from which she has harvested her crop of nectar, recruiting additional foragers. Biologist Fred Dyer sets the scene:

A successful forager returns home from a rich food source and is greeted by other workers who, if she is carrying nectar, induce her to regurgitate her load to them. If this welcome is enthusiastic enough, the forager begins dancing on the vertical sheet of comb. The dance consists of a series of repeated waggling runs in which the bee moves in a particular direction along the comb while waggling her body from side to side. During the waggling run she also emits a burst of sound by buzzing her wings. After each waggling run, the dancer circles around and realigns herself to begin the next waggling run. As the bee dances, she is encircled by 1–6 other bees that face toward the dancer and follow her movements. The dance followers observe several waggling runs and then leave the nest. Many of these eventually reach the same feeding place that the dancer had found or a feeding place close by. (Dyer 2002, pp. 918–19; see Figure 22.1)

The standard account identifies three kinds of information encoded in the dance: the profitability or value of the source, its distance from the hive, and its direction from the hive. The *profitability* of the source is proportional to the number of waggle runs in a complete dance, that is, dance duration. The *distance* to the source is correlated with the length of each waggle run along the area of the honeycomb devoted to dancing (the "dance floor") before the dancer circles back, as well as other aspects proportionate to this length, discussed later. The *direction* to the source undergoes a more complicated translation: The dancer orients its waggles at an angle diverging from straight up the honeycomb that approximates the angle of divergence between the source and the axis running from the hive to the sun's position on the horizon, or azimuth (see Figure 22.2).

In the decades after von Frisch's work, these extraordinary correlations between dance and food source directed many researchers toward questions that seemed almost ideational, assigning to bees symbols and language-like processes. How do foragers internally encode the information they bring back? How do they translate this information into gestures of the dance? How do followers of the dance retranslate its gestures into information useful in their own foraging? Describing the waggle dance as a *symbolic* activity, in particular, remains today an entrenched commonplace, repeated even in accounts where it is otherwise implicitly or explicitly contradicted (for example, see Seeley 1995, p. 36).

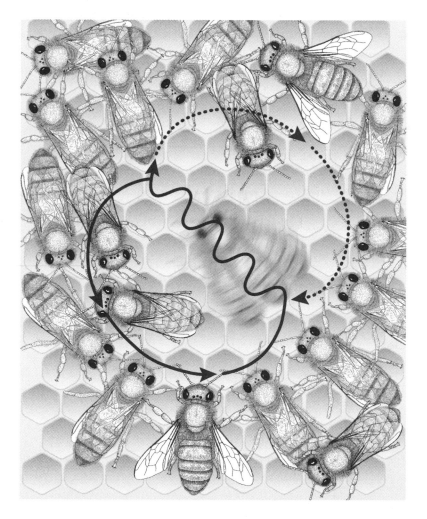

Figure 22.1. Honeybee waggle dance, waggling dancer center. After each waggle run, the dancer circles back to the start, alternating directions of the return (solid and dotted arrows). Drawn by Virge Kask after a photograph by Christoph Grüter.

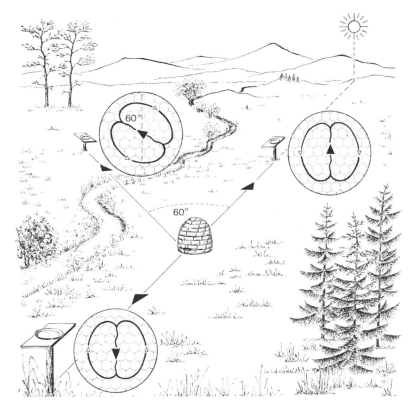

Figure 22.2. Direction of waggle on the honeycomb, as correlated with sun azimuth for three sources. The upper-left source results in a dance oriented 60° from vertical on the comb, the lower left, 180° from the position of the sun, in a dance straight down the comb. Reprinted from Friedrich G. Barth, *Insects and Flowers: The Biology of a Partnership*, 1982, with permission of Princeton University Press.

Research since the 1980s has gradually led away from such linguocentrism and toward views of dance communication as intricately evolved information channels — channels, we will see, of causal and not semantic information. Nowhere is this clearer than in the explanation of the first correlation of dance and food source, between duration and profitability. Here we encounter what Dyer has called "a well-calibrated network of social feedback mechanisms" (2002,

p. 930). We can follow its intricacies by summarizing Seeley's extended account (1995, chaps. 5–6).

The number of waggle circuits a forager dances — the aspect of the dance correlated with the profitability of the source — increases with the quickness with which she finds, on returning to the hive, food storers to offload her harvest. The offloading is accomplished by her regurgitation of her load to the storers, called *trophallaxis*. Longer wait-time for offloading results in fewer waggle runs or no waggle dance at all. The wait-time, however, does not depend on any information conveyed by the forager to the food storers about her source or load, but instead on other factors: the number of other returning foragers also needing offloading and the amount of honey already stored in the comb, which alters proportionately the time and labor needed for the storage of new nectar. The negative feedback cycle in partial view here — more stored nectar and more returning foragers resulting in slower offloading, resulting in turn in shorter waggle dances or none at all — is completed by the slowdown of recruitment of new foragers resulting from fewer waggle dances. This leads to less incoming nectar and so to quicker offloading of returning foragers.

Seeley describes another feedback cycle linked reciprocally to this one. As wait time for offloading increases for any returning forager, her probability of waggle dancing plummets, and the probability of her engaging in a separate behavior, the *tremble dance*, rises. She will perform this long-term trembling motion — averaging almost a half hour, as opposed to waggle dances usually lasting less than a minute — moving through the hive, not restricting herself to the dance floor just inside its entrance. This removes her from the offloading place, and her load will not be regurgitated until she stops trembling, thus slowing the influx of nectar — an adjustment the need of which was signaled by the slowness of offloading that stimulated the tremble dance in the first place. Her movement through the hive probably spreads chemical signals that inhibit recruitment of foragers. (We have already seen that returning foragers circulate the pheromone EO, which stimulates workers to feed the brood and increases the

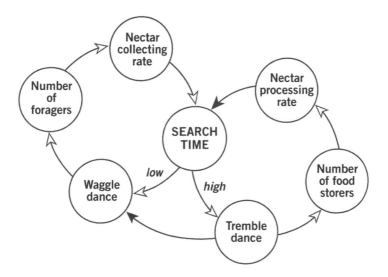

Figure 22.3. Linked feedback loops of waggle dance and tremble dance, as analyzed in Seeley 1995. White arrows represent excitatory effects, black arrows inhibitory ones. Short search times for offloading lead to more waggle dances, foragers, and nectar collection; long times lead to more tremble dances, food storers, and nectar processing. High rate of nectar processing decreases foragers' search time for offloaders, while high nectar collecting increases it. Sound bursts from tremblers inhibit waggle dancing. Drawn by Virge Kask after Seeley 1995, p. 172.

average age at which they become foragers. The tremble dance may be an important means of this circulation.) While still trembling on the dance floor, also, she produces with her wings brief bursts of sound that inhibit waggle dancing of other bees. In several ways, then — through chemical and sonic signals as well as in her unavailability for offloading — a forager's tremble dance acts as negative feedback in response to the food storers' signal of abundant nectar. The tandem balance of waggle-dance and tremble-dance feedback cycles is represented in Figure 22.3.

On the dance floor, what is conveyed to a follower of a waggle dance about the quality of the source involved? The short answer from Seeley's findings is *nothing.* An unemployed forager does not view whole dances on the floor but at most six to twelve waggle runs before she takes off in search of a source, so she cannot interpret the total number of runs in a dance as a gauge of source quality. More often than not her foraging effort is unrewarded, and she returns to the hive empty-handed to feed and resume her following of dances. Eventually one of her forays proves successful — this may or may not be because she found the source of the forager whose dance she followed — and she returns with a load of nectar. Then she joins in the dynamic of wait-time for offloading.

If nothing is conveyed to a follower about source quality, why reflect it with increased numbers of waggle runs in the dance? In fact, the question is misleading, suggesting a content where there is none. Rather than reflecting source quality, longer dances afford more opportunity than short dances for other bees to follow them, resulting in more recruits. The impact of increased dance length is a demographic one, causal information spread across the whole population of potential foragers on the dance floor, not semantic information conveyed to single or multiple followers. A forager herself integrates information concerning her food source that shapes her dance, as Seeley makes clear, and we will return to these integrative cognitive capacities. But this is not tantamount to sign formation, and in waggle dances it results in no content being imparted to dance followers.

If a dancer reflects the quality of the source in any way that followers might register, she might do so by dancing especially vigorously, though evidence of this and of recruits' differential response to it has been difficult to collect (Seeley, Mikheyev, and Pagano 2000; Waddington 1982, 2001; Hrncir et al. 2011). Vigorous dancing might be a physiological response to a straightforward chemical signal: high concentration of sucrose in the collected nectar. The concentration of this fundamental resource is a basic signal for honeybees. We have seen that the pheromones QMP and BP both can affect the threshold of sucrose responsiveness in workers, with effects on their transformation to foragers, and we will describe other effects of the signal and a neural mechanism regulating it (see also De Marco, Gil, and Farina

2005 and Abou-Shaara 2014). Vigorous dancing, like the tremble dance, might increase the dispersion of volatile chemicals on the dance floor, including pheromones and floral scents carried from the source.

All this amounts to a complex provisioning economy of the hive in all its transformative phases, involving the whole worker population — nurses, foragers, and so on. This population may number many thousands of bees; at any given moment of seasonal, daytime food collection, a hive might include several thousand foragers alone. In the provisioning economy the idea of a forager reporting back on the quality of a particular source has given way to something very different: a set of mechanisms in the homeostasis of the superorganism dispositive, connecting out to broader networks that regulate the continuum from individual bees to niche. In this aspect of the dance there is no encoded content about source profitability, indeed no translation of any code, only the interactions of foragers and in-hive workers in the feedback loops of the economy. We will return to what may be communicated in the vigor of the waggle dance, but in any case it is time to dispense with the idea of a "quality report" from the forager and replace it with yet another wonder of evolved hypermediation.

A similar causal informational approach has reshaped thinking about another use of the waggle dance, unrelated to foraging, to locate a nest site for a new colony (see Seeley 1985; Seeley and Buhrman 1999; Leadbeater and Chittka 2007). When a group of workers swarm outside the hive in preparation to form a new hive, scout bees fly off in search of appropriate nest sites. Each scout's careful exploration of a potential site is another of the marvels of programmed bee behavior. Returning to the swarm, a scout performs a waggle dance on a sheet of bees, conveying information about the site she has located. Hypotheses about the evolution of the waggle dance suggest that this might have been the earliest form of location information encoded in the dance, before it was transferred to food sources (Beekman et al. 2008). From a number of these dances by different scouts, reflecting several potential sites over a period typically lasting several days, a single site is settled upon. The swarm, led by the scouts, moves to it and establishes its new nest (Seeley 1985).

The whole process is frequently described as group decision making, but this distorts what happens across the days of dancing. In its dance, a scout conveys information about the location of its potential nest site, and other scouts are recruited to fly off and investigate, just as in foraging. Many of them will end up exploring the same site. On their return they might dance themselves, joining in the indication of that site and increasing the number of scouts dancing for it. When enough scouts do so, one of the sites comes to monopolize most of the dancers (Seeley and Buhrman 1999). Whether they do so or not depends on the persistence of the original dancer for that site, since "prolonged dances" make it "simply more likely, by probabilistic processes, that an information-seeking bee 'bumps into' a scout" dancing for that site and, after its own scouting, takes up a similar dance (Leadbeater and Chittka 2007, R710). Settling on a site involves a play of probabilities that eventually tilts toward it — a self-organizing concurrence of the scouts, with a large stochastic component but without the cognitive processes implied by the idea of a decision, let alone a group consensus of any humanlike kind.

The only place for a value judgment about competing sites to enter this selection is the idea, often asserted, that scouts dance longer for better sites. But this observation suggests some comparative conception or information about multiple sites that a single scout is not in a position to have. All each scout can do is report on a viable potential site, and we need to investigate other factors than evaluation that might determine the extension of a scout's dance. It might, for example, reflect relatively modest energy expended in her search and exploration of her site, an economizing potentially advantageous to the whole swarm.

The second and third kinds of information conveyed in the waggle dance, the spatial information about distance and direction to a source, do not evanesce on closer scrutiny the way value judgments about food sources or nest sites do. Nevertheless, as we have come to understand the mechanisms behind these kinds of information, they too have appeared to exploit causal, not semantic information, encoded and conveyed not through anything resembling symbols or language but through mediations in the superorganism dispositive.

The distance to the source is correlated with the length of the repeated waggle run, along with several other length-related aspects: the durations in time of the run and of the sound the dancer produces during the run, the number of abdomen waggles per run, and the length of the whole dance circuit (Dyer 2002). A foraging honeybee gauges distance by registering *optical flow*, the apparent motion of landmarks past it as it flies (Esch and Burns 1995, 1996; Srinivasan et al. 2000). This can be thought of in terms of the speed of that motion or the quantity of image motion experienced visually on the flight to the source (Si, Srinivasan, and Zhang 2003). Gauging optical flow is an innate capacity of many flying insects, necessary for their navigation (Graham 2010). Its neural mechanism is not completely known but might involve differential firing of neurons attuned to a range of different velocities (Ibbotson 2001). In honeybees optical flow is well developed, robust in the face of changes in contrast levels in the environment and sensed even when the moving landscape impinges on only a small portion of the visual field (for example, from below; Si, Srinivasan, and Zhang 2003). This robustness means that the mechanism can be "fooled" in changing circumstances, as experiments calibrating distances flown to durations of waggle runs have shown. Bees flying higher from the ground, hence experiencing slower apparent motion of objects below, signal shorter distance in their dances than those flying the same distance closer to the ground, while bees flying through tunnels with quick contrasts on the walls signal longer distances than they have traveled (Esch and Burns 1995; Srinivasan et al. 2000; Chittka 2004).

Translating the sense of distance gained from optical flow into the length of a waggle run involves a large scalar shift from a flight that might last several minutes to a waggle lasting less than a second, and this seems to operate according to an innate, constant relation

(Srinivasan et al. 2000; Labhart and Meyer 2002). Though the process is not fully understood, it is straightforward at least in not involving a transduction of signal, since in each case length is measured, whether in a neural registering of it according to sensory input or a behavioral response to this as output. The general neurophysiological mechanisms by which sensory input to neural nets is translated into proportionate, responsive output are ubiquitous in animals. The functional generality of such sensory-motor routines might help to explain the several dimensions of the length signal encoded in the dance — the number of abdomen waggles during a run and the duration of wing buzzing as well as length of the run. These seem to be "essentially redundant" (Dyer 2002, p. 927), and the redundancy might be an expression along several behavioral axes of the domain generality of a fundamental neural mechanism.

If in a general way we can model, in such a sensory input → behavioral output loop, a causal mechanism determining the duration of various dance gestures, we can likewise model, exploiting the same mechanisms in reverse, a decoding of the gestures by dance followers (Dyer 2002). They would reverse the scalar shift of the dancer, translating their own sensory input on the dance floor to their behavioral output in a flight covering the distance approximately signaled. As we will see, such decoding might also call on a multimodal set of sensory stimuli — tactile, auditory, olfactory, and mechanical — to gain more information about the presence and location of a food source. This is, of course, a very general account of the coding and decoding, from sensory receptors via neuronal pathways to behavior in both foragers and followers. But the model relies on transmission of sheer causal information, with relations along interconnected information channels with scalar shifts somehow built in. There is no need for semantic communication.

The same is true of the mechanisms by which the direction to the source is translated into dance orientation. The dancer diverges from straight up, as she waggles in the darkness of the hive on the vertical honeycomb, by about the same angle as the food source diverges from the axis between the hive and the sun (Figure 22.2). To do this she must gauge the position of the source relative to the sun as she flies toward it and translate this into a position relative to the pull of gravity. Unlike the case of the distance to the source, here there is a kind of signal transduction involved, employing neural mechanisms responsive to both azimuth and gravitational pull.

Celestial or sun compasses, by which the position of the sun is tracked, are widespread among animals, connected to their innate circadian clocks, and much studied in bees, ants, locusts, and a few other insects. The tracking of the sun's azimuth is a complex affair, since its apparent motion is slower when the sun is near the horizon (early morning or late afternoon) than when it approaches its zenith. Honeybees' sun positioning, as Dyer and Dickinson (1996) discovered, depends on an innate "sun-template," a mechanism that provides a default pattern for the shifting azimuth divided into morning and afternoon phases, with a relatively quick transition between them. Bees generally orient their dances according to either the morning or the afternoon azimuth, while experienced foragers can refine this mechanism further and orient their flights more precisely by learning about the sun's motions in their foraging trips. They gain this experience by locating the sun in several ways: by viewing it, by sensing the general color gradient of the sky, and, if the sun is behind partial clouds, by locating it through the polarization of its light in clear patches. The mechanism of polarization sensitivity is again widespread in insects and understood in several of them. It involves neurons that fire selectively in response to specific angles of polarization as received by photoreceptors in the insect's eyes (Graham 2010; Evangelista et al. 2014). Honeybees can derive the position of the sun from polarization vectors alone, without other cues (Evangelista et al. 2014).

Orientation in relation to gravity is another basic, innate capacity in insects and other animals, which in bees involves tiny bristles between the segments of the body and of the legs (Dyer 2002). The mechanism of transduction of azimuthal to gravitational orientation — from flight experience of the sun to dance direction — is not known, but again, as with the scalar translation of flight distance into waggle-run length, the most basic integrative capacities of advanced neural systems must be involved, crossing modalities of innate experience while converting sensory input into behavioral output. (We will return to these integrative capacities in the honeybee brain.) Evolutionary hypotheses extrapolated from simpler dances in other bee species suggest that the gravity vector might be an evolved extension of dance orientations toward visible landmarks in open-nesting bees (Dyer 2002; Grüter and Farina 2008); in honeybees in closed hives, that is, the sensation of gravity on the comb came to stand in for visual landmarks as a reference point. Darwin would have relished working out models to solve this selective puzzle.

For the dance followers to decode this directional information, another reversal of the mechanisms involved in the dancer's encoding is required. This seems on the face of it more complex than decoding distance, since it involves, again, a transduction of signal or crossed modalities. But the answer may be simpler than it seems. The followers may use moments in which they are aligned behind the dancer with her trajectory to encode their own trajectory, then reproduce it approximately in the direction of their foraging flights. There is evidence that bees that have been aligned in this way are more effectively recruited than others (Dyer 2002, pp. 929–30). We will see that there might be several ways in which followers could achieve alignment with the dancer in the darkened hive.

In all these ways, the spatial information embodied in the waggle dance relies on magnificent, innate data processing capacities by which a bee locates itself in its world, capacities that are widespread in other versions among insects and beyond. Both the encoding and the decoding of dance information can be modeled as turnaround in neural nets acting on input to shape stereotyped behavioral output—functions likewise widespread and witnessed in countless ways across all animals with sense organs, even those much simpler than a bee's.

This does not, however, complete the revisionist account of the waggle dance that recent research has offered, and additional findings bolster in several ways the case that it is a product of hypermediated but nonsemantic information. In considering the transfer of information in the dance, Dyer includes stimuli that show no sign of symbolism or language-like communication. Followers' bodily alignment with the orientation of the dancer might align their subsequent flights, as I have noted, and information may be transmitted in the sounds and vibrations produced in the dance as well as in tactile stimuli resulting from physical contact between followers and dancers (Dyer 2002, p. 928). Dyer does not include chemosensory stimuli in his list, perhaps reflecting his wariness concerning the long-running debate over the "olfactory search hypothesis," whose proponents from the 1960s to the 1990s dismissed the importance of spatial information in the waggle dance. Nevertheless, von Frisch's work already recognized the importance of odors as informational cues in and around the dance (von Frisch 1967), and they have loomed much larger in recent research — an unsurprising development in the light of new understandings of the sociochemistry of the superorganism. From an evolutionary vantage, meanwhile, the leading scenarios for the emergence of the waggle dance, working from simpler foraging communication in other bees, assign an ancestral role to odors, prior to spatial signals (Dornhaus and Chittka 1999; Price and Grüter 2015).

Chemosensory cues carried by the returning honeybee forager include the scents of flowers she has visited and food she carries (Díaz, Grüter, and Farina 2007). These odors are dispersed by the motions of the dance — the rapid waggling of the dancers' abdomen and buzzing of her wings — which propagate them especially in the air flow behind her, where followers tend to align themselves (Michelsen 2003; see Figure 22.1). More vigorous dances disperse the odors more widely and may reflect heightened arousal caused by high sucrose concentration in the nectar harvested (Gil and De Marco 2005). The alignment of followers around the dancer is additionally responsive to the kind of harvest she carries: If it is mainly pollen, loaded on her hind legs, followers tend to congregate there; if it is nectar, they gather more at her head and proboscis, where regurgitation of her

239

load to food storers — trophallaxis — will occur (Díaz, Grüter, and Farina 2007). Trophallaxis itself is a powerful spreader of chemical information about food sources through the hive. It has been shown to induce quick learning of associations of scent with food among new foragers and to activate long-term memories in experienced ones, and it is more frequently offered by foragers from sites with high sucrose concentration (Gil and De Marco 2005; Farina, Grüter, and Díaz 2005).

Returning foragers that dance also produce several organic compounds in much heightened amounts compared to foragers from the same sources that do not dance. These pheromone-like, "behaviorally active chemicals" act to increase the numbers of bees leaving the hive (Thom et al. 2007). This is a classic instance of chemical regulation of the superorganism, like the ethyl oleate also released by returning foragers, and it shows once more the intricacy and variety of the balances involved. As EO tends to increase the number of workers devoted to brood feeding, these other chemicals increase the population of foragers and may additionally attract more receiver workers to offload the dancer, signaling good foraging conditions to experienced foragers. More vigorous dances disperse more of the chemicals, and more offloading brings about more vigorous dances. A positive feedback regulatory circuit is completed.

One of the important functions of the waggle dance emphasized in recent studies is the reactivation of experienced foragers. In general, increased vigor of the dance seems to signal a heightened arousal level to the "old hands" among its followers, helping in their reactivation (Hrncir et al. 2011; De Marco, Gil, and Farina 2005). Given the many chemical signals sent by the dance, this is probably a chemosensory transfer of information — not an evaluation communicated but an automated physiological response to the composition of the harvested nectar, involving both dancer and experienced foragers. When the old hands are thus reactivated, they tend to follow few repetitions of the waggle run, leaving the hive more quickly than new recruits. Strikingly, most of them do not exploit in their subsequent foraging the spatial information offered in the dance, but instead follow their own "private" information, long-term memories of sites they have visited, newly triggered by the odors of the dance (Grüter, Balbuena, and Farina 2008; Grüter and Farina 2008).

This last finding and, in general, the increased importance of che-mosensory cues in our understanding of the dance do not invalidate the spatial information conveyed. Recent research has not simply revived the olfactory search hypothesis of the old, anti–von Frisch camp. Instead, it has underscored the "multicomponent" nature (Grüter and Farina 2008) of the informational channels involved, a complexity that reduces the centrality of spatial information without diminishing its astounding evolved mechanisms. The spatial information is now thought to function mainly as a kind of default, operating with special effect where more direct information is unavailable — where, that is, reactivated foragers do not have ample private memories of their own sources (Grüter, Balbuena, and Farina 2008) or where sources are scat-tered, short-lived, seasonally rare, or otherwise unreliable (Dornhaus and Chittka 2004; Price and Grüter 2015). Spatial information fosters "the ability to rapidly exploit the best patches when patch quality is variable," and this may be its singular benefit (Beekman and Lew 2008, p. 260; Díaz, Grüter, and Farina 2007).

At the same time as it thus circumscribes the benefits of spatial information, the revised view of the dance connects it out to the broadest patterns of superorganism regulation: the hypermediated informational systems examined in Section 7. It is in the nature of such interlooped systems that locating a starting point or original cause will be difficult and, usually, pointless. In the waggle dance, appreciation of this difficulty has grown along with our enriched understanding of the larger systemic patterns of which it is a part. In all its many signals, the dance is a miracle of emergent complexity resulting from patterns of homeostatic regulation in the superor-ganism. This, however, does not amount to the formation of inter-pretants and signs. It falls within the realm of nonsemantic infor-mational networking fundamental to all life-forms and governing, without meaning, most of them.

TWENTY-THREE

What a Bee Brain Can't Do

In 2009 Lars Chittka and Jeremy Niven took the honeybee brain as the starting point for an exercise in comparative cognition aiming to answer the question, "Are Bigger Brains Better?" (Chittka and Niven 2009). They answered with a qualified "no," but the devil was in the qualifications — and also in the question itself, which they knew invoked a tendentious debate based on an unsubtle criterion. From an evolutionary standpoint, it is hard to take seriously the ranking of better and worse brains, given the feedback relations in which all neural systems have evolved in tandem with their niches. We might more reasonably suppose that all animals' brains are *ideally* suited to the conditions of their lifeways. A monkey's brain and a bee's are both well-fitted cogs in larger systems produced through the abstract machines of evolution.

Instead of focusing on size alone, Chittka and Niven compared brain architectures, processing pathways, and capacities in small and large brains, and from this they derived several principles. First, increases in numbers of neurons devoted to processing sensory input tend to increase the detail and resolution of sensations but not to change them in wholesale fashion; rather than something categorically different, they offer increased granularity in a generally similar percept — "more of the same" (2009, p. R996). This increase, however, can involve trade-offs in performance, since elaborated sensory systems demand more information processing and reduce overall processing speed. Animal motor routines, in addition to sensory processing, also follow this more-of-the-same principle. In broad architectural terms the insect motor system is structured like the vertebrate system, in that both involve neural connections from brain to "central pattern generators," which in turn create "rhythmic motor patterns" in musculature (p. R999). In small organisms with small muscles, this architecture can be built with few neurons. The flight sequence of the desert locust, for example, requires thirty-four muscles — a large number — but only seventy-two neurons to inner-vate them, and a mere *three* neurons to form its pattern generator, firing the downstream neurons that set muscles in motion.

This reveals an especially important principle: Complex capaci-ties and behaviors can be generated from strikingly few networked neurons. This holds for miniature neural nets like the locust's flight system as well as whole, small brains. The honeybee brain (Figure 23.1) is composed of somewhat fewer than a million neurons, about 1/100,000th of the number in the human brain and two or three orders of magnitude below the numbers involved in bird brains (ranging from about 100 million to over three billion), yet Chittka and Niven can offer an ethogram of fifty-nine complex behaviors of worker honeybees. A small network of neurons can also sponsor cognitive functions such as learning and memory, as modelers of them have suggested, and in another list Chittka and Niven give two dozen learning routines that insects have been shown to enact, most of them experimentally demonstrated in the honeybee. We will return to honeybee learning, where we will see that this principle, joined to computational modeling and new knowledge of bees' brains and sensory-motor systems, can explain imposing insect achievements without recourse to humanlike or even birdlike cognition.

What about those bedeviling qualifications? They arise when Chittka and Niven consider the knock-on effects of brain enlarge-ment. Since the time and energy required for neural processes increase with length of neural transmission, which in turn increases with brain size, large brains tend to be organized differently from small ones, proliferating short-range connections and minimizing

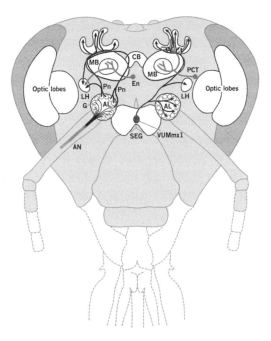

Figure 23.1. Frontal schematic view of the main nuclei in the honeybee brain, highlighting olfactory and sucrose-activated pathways. MB: mushroom body; CB: central body; LH: lateral horn; AL: antennal lobe; G: glomeruli within AL; AN: antennal nerve; SEG: subesophageal ganglion; Pn, En, and PCT: other neuronal connections. VUMmx1: sugar-activated neuron, projecting to ALs, LHs, and MBs. Pathways not shown also convey visual stimuli from the optic lobes to the MBs. Pathways are bilateral, though shown here on one side only. Drawn by Virge Kask after Barron et al. 2015.

long-range ones. This promotes the segregation and multiplication of distinct brain areas rich in local connections, a circuit diversification and modularity that can sponsor novel capacities and behaviors. The song system of songbirds, with many nuclei rich in internal connectivity and relatively small numbers of neurons connecting to other nuclei, offers an example of Chittka and Niven's effect. In the insect brain, modularity can be seen in the *mushroom bodies (corpora pedunculata)*, the centers of learning, memory, and informational integration (Figure 23.1). Insects with large mushroom bodies such as honeybees show greater distinction of areas within them than simpler insects. Responding to sense-organ input, these areas mediate sensory processing and facilitate cross-modal integration between, for example, olfactory and visual percepts (pp. R1004-5). Nevertheless, the variety of distinct modules in the most complex mushroom bodies doesn't approach that of vertebrate systems like the song system.

Local, modular connectivity is probably correlated with memory capacities, especially the storage of long-term memories, which in many vertebrates far outstrips anything found in insects. We saw in Section 14 that the hippocampus is associated with long-term memory in mammals, and that its connections to cortical processing are mediated through parahippocampal nuclei proximate to it; analogous structures seem to be at work in bird memory. Increased storage could result from the parallel processing enabled by the proliferation of short-range connections in these distinct brain regions, which might lay down memories in multiple circuits. Working memory, Chittka and Niven note, also might benefit from such storage, since "searching a bigger library of stored information might enable an animal to generate more and better solutions to a problem" (p. R1004).

This points toward one more, large qualification, a very general cognitive difference between large and small brains. Expanded parallel processing and more numerous stages of processing across serially connected modules foster in vertebrates a degree of cross-modal integration of information not witnessed in insects. This enables "the computation of variables not directly represented in sensory inputs"; new, different kinds of information can be extracted from input, giving rise to flexible responses to changing stimuli and "novel receptive fields" (p. R1005). Chittka and Niven's example is avian: barn owls' ability to create spatial maps from auditory input. The songbird song system forms another such multiply parallel, serially extended network.

Bigger brains, then, are not better — this is a non sequitur in the light of evolutionary machines — but neither are they nothing but more of the same. There comes a threshold, or perhaps there come many thresholds, beyond which quantitative increase results in qualitative difference of function and capacity. As layers and interconnections of processing proliferate, new computational dimensions, new integration, and new knowledge emerge. Affordance horizons of niche construction shift, introducing for many animals, for example, new opportunities in social affordance. The contextual experience of the world thickens with the appearance of episodic memory, an especially complex kind of long-term memory, and as it does so metarelational cognition comes into reach with its interpretant formation. Semiotic information and meaning come to be overlaid on the broad, ubiquitous transmission of causal information. Songbirds sit firmly ensconced in this realm of aboutness, but, from the physiological, neuronal, and cognitive vantage of Chittka and Niven, it seems that honeybees do not — even though they represent an apex of invertebrate complexity. Let's examine their cognitive capacities more closely.

Since the 1990s, behavioral and cognitive studies of honeybees have shown that they exploit both modular and integrative processing (see Menzel and Giurfa 2001). The modules are innate, domain-specific neural circuits that respond to particular input and automatically produce physiological change and behavioral output. They can function vertically, independent of other modules, creating discrete sensory-motor loops such as those discussed earlier: responses to pheromones, measure of distance through optical flow, and the locating of azimuth dependent on the sun compass template. Some of these modules are sometimes called "value systems" since they signal specific affordances of high importance to honeybee subsistence and induce behavioral responses to them. (But remember that the value here mooted is not a consideration, judgment, or evaluation; it is our measure of centrality of an affordance in bee lifeways.)

One of the most prominent and best understood of these systems is composed of a single, long neuron especially sensitive to sucrose concentration (Menzel and Giurfa 2001; Hammer 1993; see VUMmx1 in Figure 23.1). It extends from the honeybee's proboscis up to the antennal lobes (the bee's first-stop olfactory processing centers) and on to higher olfactory centers in the mushroom bodies and to

behavioral output centers (the lateral horns). The compound associated with synaptic transmission from this neuron to these areas
is octopamine, a hormone connected also to heightened arousal in
invertebrates (Giurfa 2007), so here we may see an automatic, modular mechanism underpinning the responses to sucrose we encountered in foraging and the waggle dance. The systemic economy of
this sensory-motor loop is breathtaking: A single neuron forms the
substrate for honeybees' sensation of food rewards and points down
pathways leading to learning, memory, and behavioral responses.

While such modules can act with vertical independence, they do
not always do so. Many studies have shown that input from the modules can be integrated horizontally in downstream cognitive processing, resulting in crossing and mixing of their domains (Menzel
and Giurfa 2001). In "delayed matching to sample" experiments, for
instance, bees are trained to locate a sucrose reward at the end of one
branch of a Y-shaped maze by matching a stimulus at the entrance
with the same stimulus at the reward. Bees that learn to associate a
visual stimulus at entrance and reward can transfer this learning to
a maze deploying an odor as the stimulus at beginning and end—a
horizontal integration of input from separate sense channels that
transfers a learned pattern of behavior (pp. 67–68).

Many honeybee behaviors seem to require such integration. Several cognitive pathways join in navigation, merging responses to
innate templates with learned and memorized landmarks and routes
(Menzel and Giurfa 2001). Horizontal processing probably also underlies the related capacity for *path integration*, by which honeybees,
ants, and other insects monitor the twists and turns of foraging journeys so as to derive from them a straight line back to their nests (Graham 2010). In honeybees, integration is suggested in the refinement of
the sun-compass template through information about the landscape
learned in exploratory foraging flights, and it may be reflected in the
signal transduction from azimuth to gravity in the waggle dance. It
certainly must play various roles in the multicomponent information
transfer that regulates foraging as a homeostatic economy, including

the dynamics of the dance. In all these cases, integration of distinct modular systems results in increased variety of response to the myriad stimuli coming from the outside world. Integration modulates automatic responses to encounter more flexibly the lived context, from society to abiotic conditions.

The mushroom bodies, centers for multimodal integration in honeybees and other insects (Chittka and Niven 2009; Giurfa 2013), are central also in memory formation and learning (Smith, Wessnitzer, and Webb 2008; Hourcade et al. 2010). In honeybees, foragers in particular rely on memories, and the varying durations of their short-, middle-, and long-term memories correlate with challenges of different foraging situations: within a patch of flowers, between patches, and in separate trips following short or long interruptions (Menzel 1999). Memory formation is intimately connected to associative learning, especially through the sucrose pathway described earlier. A short-term memory, lasting a matter of seconds, can be formed in response to a single reward, a long-term one, lasting a worker's lifetime (usually a few weeks), after a few rewards. The neural substrates of the long-term, stable memories have been traced to increased synaptic density in portions of the mushroom bodies dedicated to olfaction, and their stability reflects stimulus-induced alterations in RNA transcription and genetic expression (Hourcade et al. 2010). There is no doubt, in short, that learned associations laid down as memories and reflected in neuronal changes underlie foragers' abilities to relocate and exploit sources of nectar discovered earlier, even when they are reactivated after long nonforaging periods. These memories form the private information of experienced foragers, mentioned earlier, that enables them to retrace their own earlier routes after experiencing another forager's waggle dance.

The associative learning here is impressive in its quick efficacy and potential for long-term stability (for a review, see Giurfa 2007). How far beyond sheer association, however, does honeybee learning extend? How complex can the behaviors it enables become? Learning feats of honeybees form an imposing list (Chittka and Niven 2009), but in explaining the mechanisms that underlie them there is a temptation to overshoot the mark and insert humanlike strategies and capacities — one more instance of anthropocentrism effacing the marvels of organisms functioning in other ways. The tendency was especially marked in the 1990s, the early, heady days of discovery of cognitive complexity in insects, but it persists today.

An example concerns honeybee *numerosity*. Many experiments have shown that honeybees can learn to discriminate between visual stimuli containing different numbers of elements and use this ability to find rewards in various settings (for a review of research, see Pahl, Si, and Zhang 2013). The ability falls off sharply for more than three elements, but within this limit it can be exploited both in simultaneous presentation of stimuli — designs of two vs. three colored circles, for example, with one design connected to a reward, the other not — and in sequential presentation, where bees are trained to seek food after a certain number of landmarks (Chittka and Geiger 1995). These findings are robust, and experiments have been carefully fashioned to filter out effects from aspects of the stimuli other than their number.

What remains mysterious is the cognitive mechanism involved, and the mystery continues to encourage strong claims concerning honeybee counting. The title of a study from 2019, for example, announced that "Numerical Cognition in Honeybees Enables Addition and Subtraction" (Howard et al. 2019). In its delayed matching to sample experiment, foragers in a Y-maze learned to find a sucrose reward when a first stimulus including several elements needed to be matched to a second with either one more or one less element, with the increase or decrease correlated with the colors of the stimuli. Thus, for a blue initial stimulus of two elements, the rewarded second stimulus consisted of three, while for a yellow, three-element initial stimulus, the correct later stimulus showed two. The bees performed correct searches in a proportion that grew linearly across many learning trials, and trained bees remembered the correct options at a statistically significant level, demonstrating associative learning capacities integrating different kinds of visual stimuli. But does this demonstration warrant the researchers' assertions that honeybees do arithmetic — that they "can learn to use blue and yellow as symbolic representations for addition and subtraction" and manipulate mental "representation[s] of numerical attributes" (pp. 1, 3)?

Another recent study of honeybee numerosity suggests not and offers a neural model by which bees might track small numbers of objects (Vasas and Chittka 2019). The model requires only four units, which may minimally be thought of as four neurons, linked in balanced stimulation and inhibition and responsive to changes in brightness from visual input. The spiking of a "brightness neuron" responding to input feeds into short-term, working memory along two paths. It maximally excites one neuron that feeds back on itself, maintaining a maximal output for every change in brightness (a consistent "brightness memory"); it weakly stimulates another neuron, which also feeds back on itself and so accumulates its output in proportion to the number of changes of light and dark areas the bee moves through (a cumulative brightness memory). Both neurons feed into a fourth neuron, with the consistent brightness memory exciting it, the cumulative memory inhibiting it. The dual effect on this fourth neuron is a phased excitation in any situation of changing brightness, such as a patterned object inspected by the bee, and this generates "a continuously updating evaluation of [its] numerosity" (p. 86).

This model accords with the nature of bee vision, which registers brightness, shapes, and their borders, and with a central aspect of bee foraging behavior: the close-up, sequential inspection of objects or features of patterned objects. It models a behavior that looks to us like counting using a mechanism with minimal components, all amply attested in bee brains without concept formation or numerical symbolism. Its "numbers" are not integers or symbolic representations at all but accumulations of magnitudes of neuronal outputs, and there is no sign of anything like human conceptualization of addition and subtraction. Vasas and Chittka explain how their simple computational networks might sponsor other tallying capacities bees have shown, such as landmark counting on their foraging journeys.

Even more significantly, they might underlie capacities in bees that are often taken to require generalized concepts or rule-based choice: responses to stimuli found in a generalized spatial relation to other prompts (for example, above or below them) and learning to respond either to a stimulus identical to one previously viewed or to one different from it (discriminating sameness and difference).

The broad question raised here concerns the potential for abstraction in honeybee cognition, as this might be manifested in the kinds of visual learning bees can achieve over and above sheer associative conditioning. For example, in the last-named learning situation, do bees abstract a general concept of sameness and difference from stimuli and then apply it to their processing of novel input?

It is well established that bees learn to follow visual stimuli in a more sophisticated way than by merely matching them to a memorized template and that they can respond to novel stimuli that only fuzzily resemble learned ones. This capacity for *generalization* seems to require them to "extract and combine the specific features of an image to create simplified sketches" applicable to other, nonidentical images (Avarguès-Weber, Mota, and Giurfa 2012, pp. 261–62) — here is the presumption of a mechanism for abstraction. But it may instead be a case of complexity built on straightforward salience: a response to individual features shared by different stimuli that are important in bee lifeways, possibly with some integration across sensed parameters such as shape, brightness, and color that would not, however, rise to the level of a generalizing concept. Honeybees' response to sucrose provides a relevant model from the olfactory domain in which a single reward network operates in diverse circumstances, from feeding at a flower to trophallaxis in the hive, and leads to many responses. No generalization is required here, let alone conceptualization, but only the spiking of the neuron at the heart of the system to generate various motor responses in different situations, sometimes joined with downstream cognitive integration dependent on both innate mechanisms and memorized experience.

Categorization, a capacity to form classes of stimuli with well-defined borders such that a stimulus either belongs to a class or is excluded, is a more advanced ability than generalization assigned to honeybees, ostensibly overleaping a higher bar of abstraction (Benard, Stach, and Giurfa 2006). Honeybees' achievements in this area are striking. They can transfer learned responses from one visual pattern to a different one that resembles it only in general aspects of its orientation or shape. Aspects that enable this categorical clumping include the orientation of edges defined by light/dark contrasts, radial and concentric designs, angular symmetry, and bilateral symmetry, and it is no doubt significant that some of these correlate with landmark patterns and natural objects central to foraging, particularly flowers, while others correlate with bees and other animals. More impressive still, bees can process multiple orientation cues of complex patterns and transfer them to novel patterns with the same set of cues. Once again, some form of abstraction is usually deemed necessary for these abilities — the idea, for example, that bees "extract regularities in their visual environment and establish correspondences," generating "a large set of object descriptions from a finite set of elements." In this way bees would answer the need for "cognitive economy" in a small brain better than by storing exact templates of every object and landmark viewed (p. 267).

Here too, however, as in the cases of numerosity and generalization, students of honeybees have pointed out that such complex "stimulus classification" might be achieved through simple neural mechanisms, without recourse to rule- or concept-governed definitions of the categories. The different orientations and shapes that are salient for bees might fire neural detectors innately attuned to them (Benard, Stach, and Giurfa 2006, p. 268), of a sort known to exist in the bees' optic lobes (Nordström and O'Carroll 2009; for the optic lobes, see Figure 23.1). Multiple activation of different kinds of detectors could enable multiple-cue recognition to be pieced together from "a combination of low-level features" (Benard, Stach, and Giurfa 2006, p. 267). To add learning into the equation, these detection systems need only be attached to a neural element associated with the reward at the end of a search, and here once again the sucrose system provides a simple case-in-point from another sensory modality. The conclusion follows that "category learning . . . could . . . be reduced in the honeybee brain to the progressive establishment . . . of an associative neural circuit relating visual-coding and reinforcement-coding neurons, similar to that underlying simple associative (e.g., Pavlovian) conditioning" (p. 268).

And so it goes: In each case where advanced conceptualization is attributed to honeybees, simpler mechanisms offer compelling explanatory models. One more kind of learning sometimes ascribed to honeybees, *context learning* (Collett, Dale, and Baron 1997), can also be reduced to a multiplication of associative connections. It involves the use of distant landmarks to prime memories of navigational routes and locate nearer, specific targets. It can be modeled as an interaction of wide-field and narrow-field receptor neurons in the optic lobes, priming memories of learned routes (p. 351), and it might also exploit the multi-cue capacities for pattern processing shown in categorization experiments. But it need not entail any fine-grained memory of a route, full-field processing of the landscape, situated memory of a whole earlier flight, or cognitive map of a territory such as that of the barn owl cited by Chittka and Niven (see Cruse and Wehner 2011; Cheung et al. 2014).

In all these kinds of learning we see impressive behavioral achievements that do not require humanlike or even widespread vertebrate abilities to explain them but can instead depend on information pathways emergent from simple neural nets. Such mechanisms have the advantage of answering well to what we know of honeybee sensory and cognitive architecture, and they take plausible account of the economies at stake in a small brain. They have proliferated in recent research (for example, see Peng and Chittka 2017 and the studies in Giurfa, Riffell, and Chittka 2020), offering alternatives to the humanist vocabulary of rules, concepts, mental representations, and the like that dominated the first years of the cognitive revolution in invertebrate studies.

TWENTY-FOUR

Meaning and Meaninglessness

In these various modes of honeybee learning we see no sign of one more cognitive function combining learning and memory in a rich interaction, one basic to animal indexicality. This is *episodic* experience, in which an animal's responses are conditioned by earlier experiences preserved in situational detail and bounded coherence — preserved as an episode in its life. Episodic experience is absent from protocols of learning that have been demonstrated in bees (see Chittka and Niven 2009) and from the two other central measures of honeybee complexity examined earlier: the regulation of society in the superorganism and the waggle dance. In these cases complex behavior is generated from the sociochemical regulation of superorganism homeostasis, the related loops of causal information of the dance, and the neural networks of associative learning, without episodic memory or its situational learning. We can say the same of subroutines within these three cases: regulation of reproduction in the hive, danced communication of direction or distance to a food source, formation of consensus for a new nest site in the swarm, foragers' retrieval of route memories, and matching of multiplex patterns to locate rewards. All these spring from the emergent action of signal chemicals, sensory mechanisms tailored to salient affordances, simple neural nets with minimal modular distinction, and the accumulation of learned associations and memories. None of them require a cognition that tailors episodes whole-cloth from the continuous fabric of sensory input.

The semiotic machine, the fourth abstract machine of evolution, springs into operation in the presence of situational learning and the episodic memory on which it depends, where arrays of input can be stored, hierarchically ordered, and retrieved as coherent wholes. Beneath these capacities stand the strongest forms of attention — not merely the generation of differential salience among incoming stimuli, which most animals, vertebrate or invertebrate, display in some form, but the voluntary focus that brings about a recursive parsing of the external world. Such attention drives the semiotic machine because interpretant formation is analytic, relating aspects of one thing to aspects of another and thereby forming an essential characteristic of the sign. The concomitant of parsing and analysis is experience of larger situations from which aspectual relations are drawn. The interpretant is inherently bound up with broader situational cognition. The metarelation is inherently episodic.

Songbirds amply enact episodic memory, situational learning, and the indexical semiosis that arises with them, as our case studies have shown. Song sparrows' countersinging reflects the experience of hierarchies flexibly determined by episodes in the earlier experiences of the two birds involved with one another and other neighboring birds, bringing them into relation with less episodic pathways toward song learning and performance. The pseudo-duet of a magpie-lark entails situational knowledge on both sides, performer and auditor, and even builds in an interpretant of an interpretant on the part of the deceiving singer. The sexy syllables of male canaries function as more than honest signals tapping innate responses, revealing a recursive parsing of experience and built-in possibility of situational learning. They engage females in comprehensive weighing of situations, evaluating male messages in ways that reflect the complex memory and learning of the song system.

Needless to say, birds also embody meaningless systems of causal information, as do all organisms, and we have traced avian molecular pathways, from external stimuli through immediate early genes to altered genetic transcription, that are similar in outline to bee pathways followed here. Indeed, these pathways can show connections stunningly conserved across evolutionary history. For example, the same immediate early gene, ZENK, that we saw inducing neural plasticity in canaries is upregulated also in foraging honeybees, fostering plastic changes in their learning and their memorization of optimal foraging times (Shah, Jain, and Brockmann 2020). The difference is this: While these and related pathways interact to provide a robust explanation of the complexities of bee behavior, they don't suffice to do so for birds. Over and above the processes at work in bird behavior that resemble determinants of bee behavior, bird cognition adds something more. Its massively parallel, lengthily serial processing in the song nuclei, connecting out to the (massively parallel) centers and (extended serial) networks forming memories and enabling learning, bring about not merely a quantitative increase in complexity of behavior but qualitative shifts as well, enabling the computation of new kinds of variables, in Chittka and Niven's language. Bigger brains with increased modularity offer different possibilities for en-niched cognition than smaller brains.

The differences between brains processing causal information to create immense behavioral complexity and brains transforming causal into semantic information reflect these differing possibilities. I have related semiosis and meaning to the cerebral processing in birds involving networks between the cortex and deeper areas: the structural loops in the song system from pallium through striatum to thalamus and back, involved in learning, memory, and attentional focus and mediating top-down signals from the forebrain and bottom-up input from the sense organs; and the networks from cortex to hippocampus, the storehouse of episodic memory. These networks are aspects of the increased hierarchization of processing that distinguishes the avian from the insect brain and, in general, larger from smaller brains.

I began this discussion of honeybees by resisting the humanist parochialism that reads symbols, meaning, decisions, and the like into bee activities, at the same time resisting the older view of insects as simple reflex mechanisms. Have I come around finally to reaffirming this older view, reducing even the magnificent honeybee to an automaton? No, for the newest understandings of honeybee behavior depend on the powerful informational causality that arises in networked, reciprocal interactions of simple elements. The honeybee dispositive is an example, perhaps unparalleled, of such power: power to generate the layered structure of the superorganism, to regulate the many dimensions of its ongoing relation to its environment, and to mold via reciprocal impacts everything from genes to niches — with waggle dances along the way. We know enough now to see that pinning such complexity to the mast of meaning represents nothing so much as our failure, in the face of eons of evolutionary machination, to extend the human imagination to the domain of meaninglessness.

PART FOUR

Outstanding Questions

What is the general status of the semiotic machine

among the processes of evolution?

How has it conduced to the advanced capacities of some animals

to form technologies and cultures?

TWENTY-FIVE

Questions Concerning

Evolution

A semiotic major transition in evolution?

Darwin's abstract machine of *selection*—inheritance with variation in a situation of limited affordance—always works in a changing environment, altered in part by the organisms in it even as it
shapes them. This feedback loop between organism and environment, change in the one mutually changing the other, constitutes the
second abstract machine of evolution, *niche construction*. Expression
of the signals by which organisms regulate their systemic identities
amid such change must be flexible and plastic, shifting with their
shifting niches. The plasticity derives from layers of *mediation* creating informational pathways out from genetic molecules to the niche
and back in again, from niche to genes: the third abstract machine.
Mediated chains of processes operate from moment to moment,
with no divide separating the quickest processes from the long-term
viability of organisms and populations of organisms in their niches.
Living systems are radically open to their environments, and niche
construction must be gauged at time scales reaching from the near-
instantaneous to the epochal.

These evolutionary machines more than sufficed to create stunning diversity and complexity across the earthly biosphere. The
honeybee superorganism arose from their machinations, amazing
us with the informational intricacies of its homeostatic economies.

Countless other organisms astonish as well—microbes, plants, and animals all embedded in their own distinctive (and radical) niche constructions. Such intricate, looped networks of causal information are a requirement for life, and they help to differentiate causal information within the biosphere from simpler causal information outside it.

After several billion years of diversification generated by these machines, a *semiotic machine* emerged in the niche construction of some organisms. This decisive event was founded on neural systems linked in multiple levels of reciprocal connectivity, which sponsored episodic memory and situational learning, with attentional focus recruiting both. Such cognition created percepts characterized by recursive analysis of parts and wholes, parsing sensory data so as to relate things to one another through aspects of each. These percepts are interpretants; they form the metarelation of the sign and bring meaning into the world. For the first time on our planet, a new kind of information carrying semantic content arose on the foundations of causal information.

The semiotic machine may or may not have brought about a quantitative shift in the transformative power of biotic information, a general increase, in other words, in the evolvability of some life-forms. But it certainly formed a new kind of transformation. Its metarelation introduced a new dynamic to the shaping of niche affordances, altering affordance horizons for the organisms that exploited it and changing the niches of all other organisms whose lifeways intersected with theirs. The emergence of semantic information marks a discontinuous outcome of far-reaching impact in the operation of the first three machines, and in this way it rises to the level of a *major transition* in the earthly history of life.

This concept was introduced by John Maynard Smith and Eörs Szathmáry in the 1990s to accommodate the growing evidence that evolution has proceeded not only in the kind of incremental change described by Darwin but also, at a number of decisive turns, in quick, dramatic innovations (Maynard Smith and Szathmáry 1995;

Szathmáry 2015). Major transitions they proposed include the origin of the cyclic chemical systems of metabolism, of information-bearing RNA and DNA molecules, of cells with distinct organelles in them, of multicellular organisms, of canalized development rendering organisms resistant to selective change, of sexual recombination in reproduction, of societies of organisms, and more. At the end of their list comes human language, deemed the essential means of differentiating human from nonhuman societies and bringing about human mind and culture.

From the semiotic perspective, we see that something has been skipped over in this list: the transition to the perceptual metarelation of the interpretant, thus to signs and meaning. Meaning did not await human language to take shape. It informs the lives of thousands of kinds of animals today, with language only one of its many outgrowths — if an especially important one in its impact.

One semiotic transition or more?
But perhaps it would make sense to speak in the plural: *transitions* to semiosis. The relation of avian to mammalian semiosis seems to be a close one, depending on some neural substrates that are homologous, but it seems also to involve other analogous, convergent features and processes. In this case we might entertain the idea that a transition to semiosis occurred twice, both in the dinosaur → bird clade and in mammals, with the two groups veering toward one another in the semiotic processes arising in their complex neural systems. This convergence, however, might not have started from a dramatic distance, since it relies on brain structures and kinds of connectivity shared by birds and mammals, some of which were presumably present in their last common ancestor, some 300 million years ago. Perhaps, then, the bird/mammal semiotic proximity represents a convergence not on semiosis in general but on highly elaborated forms of it, starting from an incipiently semiotic common ancestor.

The possibility of a more dramatic convergence remains, however, because of the striking, outlying case of cephalopods: octopuses, cuttlefish, and squids. Here it is implausible to posit a common ancestor with any semiotic capacities. The last link of mollusks and vertebrates is hard to track through fossil evidence, given the soft-tissue nature of the organisms involved, but it was long assumed to sit at least as far back as the Cambrian period, half a billion years ago. In the 1990s, new techniques for timing the divergence of genetic sequences between these groups doubled the estimate, placing the last common ancestor at least a billion years ago, far back in the Proterozoic period and deep in the presemiotic stage of the earthly biosphere (Wray, Levinton, and Shapiro 1996). If cephalopods form interpretants and signs, then, they would manifest a convergence toward mammals and birds from a huge phylogenetic distance — two utterly independent transitions to semiosis.

This is a big "if." Cephalopods are creatures of varied and fascinating behavior, with central nervous systems of far greater complexity than those of any other mollusks (Darmaillacq, Dickel, and Mathers 2014). Their famous visual signaling, changing their skin colors and flashing patterns of contrast and iridescence along their bodies, ranks among the most intricate and beautiful of animal behaviors, and like the honeybee waggle dance it has regularly inspired talk of cephalopod language and symbolism. It works under direct brain control, exploiting expansion or contraction of tiny muscular organs containing sacs of pigment (chromatophores), which can number in the millions. These combine with iridescent and white reflector cells deeper in the epidermis to create the patterns (Hanlon and Messenger 2018, chaps. 2–3; Mäthger et al. 2009). The patterns are employed to alarm or startle predators or as camouflage from them and in intraspecies social communication, especially in sexual interactions, where they signal receptivity and ward off rivals. The intricacies of such communication can be striking, as when a male cuttlefish signals its courtship of a receptive female with a pattern along one side of its body while simultaneously deterring a male rival with a different pattern on the other side (Brown, Garwood, and Williamson 2012). Skin-pattern communication is combined with a variety of bodily postures to form multimodal displays, adding to the informational complexity (Hanlon and Messenger 2018, chap. 7).

Even this does not reach the limit of cephalopod capacities. Experiments have shown many species of cephalopods, especially among octopuses and cuttlefish, to be capable of various kinds of learning, simple habituation and sensitization to repeated stimuli as well as associative learning of impressive intricacy (Hanlon and Messenger 2018, chap. 8). This includes discrimination of stimuli presented simultaneously or even serially, and it can be induced visually, through touch, or in an integration of the two. Spatial learning also is well documented in octopuses, which can take twisting foraging routes away from their dens but swim straight back to them when interrupted or threatened. Foundational for such learning is memory, and octopuses have been shown to have independent systems of short- and long-term memories with some evidence of distinct brain localizations of the two (Hanlon and Messenger 2018, chap. 8). Distinct formations of short- and long-term memories are also suggested in cuttlefish experiments (Messenger 1977; Dickel, Chichery, and Chichery 1998). One experiment has even been offered as support for the formation of episodic-like, what-where-when memories in the common cuttlefish *Sepia officinalis* (Jozet-Alves, Bertin, and Clayton 2013).

Does all this reveal the formation of interpretants and signs, and so a transition to semiosis independent of that of birds and mammals, or is it a fascinating case of highly elaborated signaling — causal and not semiotic information? This remains unclear at the present state of the evidence. Evolved animal behaviors concerning reproduction, predation, and defense are mind-bogglingly varied and highly complex, even in animals without plausible claims to semiotic cognition. The case of honeybees shows that astonishing social and communicative complexity can arise from causal information alone, without signs or meaning. The visual communication of cephalopods could be another, equally astonishing instance.

The parallels of the two cases are telling. Associative learning and the generation of both long- and short-term memories are at least as well attested in bees as in cephalopods. In bees learning rises to a level of intricacy, involving cognitive integration of stimuli from more than one sense with innate templates and tendencies, that has seemed to observers to betoken other abilities than it does — abstract numerosity, generalization, categorization, and the like. Honeybees' ability to find a direct path back to the hive after a meandering foraging trip mirrors exactly octopuses' ability. It relies not on especially advanced learning but on memory and a path integration capacity that is understood well enough for us to know that it is widespread among insects. In fact, innate gauges of time (such as the sun compass), distance (optical flow registering), the pull of gravity, and other environmental features combine with multiple sensory cues to enable navigation of the niche in many, probably most animals. The experiments putatively supporting episodic-like memory in cuttlefish could well reveal just this kind of integration, wherein physiological timing mechanisms join with associative learning concerning what prey might be found where to mimic episodic memory — when nothing is present comparable to the full, nuanced life episodes stored and retrieved by a songbird interacting across seasons with rivals and mates.

What is needed to resolve the question of cephalopod semiosis is the kind of evidence that has enabled us to distinguish bees and songbirds: longitudinal evidence of behaviors in lifeways observed lengthily and fully, for example of communities of songbirds across generations and of the maintenance and reproduction of the honeybee superorganism. For almost all the seven hundred and more species of squids, octopuses, and cuttlefish, gathering such evidence poses challenges far greater than those involved in studying bees and birds.

If such evidence could be had, and if it suggests that cephalopods have converged on semiosis, this would raise the intriguing possibility that the emergence of signs is highly likely or even inevitable as neural systems attain a certain threshold of organizational and connective complexity. In general, a correlation seems likely between neural complexity in animals and features such as rich episodic memory, situational learning, and a heightening of flexible, controlled attention. If these are indeed foundations for semiosis, as I have argued, then a surpassing of a certain cognitive threshold might bring about sign-making wherever it occurs. Perhaps research now in progress on major transitions within the evolution of cognition, charting the emergence of several kinds of organizational complexity in neural systems, will help to define this threshold (see Barron et al. 2020–23).

What is probably not inevitable in a biosphere, on the other hand, is the generation of such complex neural substrates in the first place. Something greater than nine-tenths of the history of earthly life passed with nothing like them, and perhaps in the coming years exobiologists will discover biospheres where they have never emerged, in the sands of Mars or the ice-girded seas of Europa or Enceladus.

Semiosis: An all-or-nothing proposition?

If cephalopod semiosis were convincingly demonstrated, it would extend the realm of signs far afield from birds and mammals, evolutionarily speaking. I have been cautious about plotting the borders of this realm in any exact or comprehensive way, but a few generalizations can be hazarded.

My analysis of the semiotic machine and the case studies of Part III are meant to indicate together a location for the frontier that separates animals reliant on causal information alone from causal + semiotic animals, with honeybees in the first camp and songbirds in the second. Given neural and behavioral similarities respectively among insects and among birds, it seems likely that each group as a whole stands on its respective side of the border. On the invertebrate side, we might suspect further that all arthropods are nonsemiotic, but the case of cephalopods should caution us, for now, against excluding semiosis from invertebrates altogether. On the vertebrate side, given the relation of birds' brains to those of mammals and the clear presence in mammals of the neural substrates that found semiosis, it seems reasonable to locate many, perhaps all, mammals inside the semiotic sphere. Individual case studies might, of course, show this to be too hasty a conclusion and call for more granular distinctions.

Of other vertebrates — reptiles, amphibians, and fish — we must be less confident. We can easily identify vertebrates that show none of the neural hallmarks required to bring signs and meaning into the world — a primitive fish such as a lamprey, for example. We can also see across vertebrates a graded appearance of these hallmarks and associated behaviors, perhaps also suggesting a graded appearance of semiosis. Comparative analyses of the formation of complex attention, visual and otherwise, indicate that birds and mammals possess its neural correlates in forms more developed than in the other vertebrate groups and exercise a more complexly situation-based focus on salient stimuli (Knudsen 2018, 2020). A similar gradation is present in the neural systems associated with episodic memory, with its potential for recursive, parsing perception (Allen and Fortin 2013). The study of this capacity far beyond humans is in its early stages (Templer and Hampton 2013), but compelling evidence for it has not yet been offered beyond some birds and a few mammals. Perhaps semiosis is a highly restricted phenomenon, absent from chameleons, frogs, and salmon and found only in some birds and mammals. My case studies are meant only to establish the existence of a border between meaning and meaninglessness; they cannot map it fully.

At stake here is not only the issue of inclusion and exclusion — who's inside, who's outside the semiotic circle — but a separate question: Can we locate a sharp border at all, or are we faced with a fuzzy one, where some animals embody an incipient, shadowy, half-formed semiosis, and the road from causal information to sign-making is crowded along its whole length with creatures increasingly semiotic? The fuzzy option seems to me unlikely. The semiotic machine is a process that emerges from certain neural structures and connections and a certain level of complexity they bring about in processing sensory input. The result is a novel kind of percept: analytic, metarelational, semiotic, meaningful. This emergent metarelation probably does not admit of halfway states; either the sign structure appears in a cognitive stance toward the world or it does not. The neo-Peircean structure of the sign helps to pinpoint the ontological requirements for this switch to be flipped on.

Even if the border itself doesn't admit of gradations, on its semiotic side evolved gradations in the complexity of sign-making are clear. It seems likely that no animal on Earth today other than *Homo sapiens* formulates in the wild the signifying layers that characterize symbolism: indexes pointing to indexes, meaning emerging from these interrelations in single indexes, and groups of indexes pointing to the world through syntactic arrangement (Deacon 2012a). This fact has habitually been taken to exemplify human exceptionalism, which it does. All too often it has been taken to confirm a sheer break between human cognition and that of all other animals, which it doesn't.

The concept of hyperindexicality, describing the nonsymbolic organization of indexes into arrays with incipient syntactic relations among their elements, helps us to discern degrees of semiotic elaboration. I have discussed elsewhere the evidence for a *hyperindexical age* in hominin evolution, before the advent of symbolism and modern language, and argued that it signals a history of semiotic elaboration in one clade, which eventually led seamlessly to full-fledged symbolism (Tomlinson 2018). The paleoanthropological consequences of understanding this gradation are large, since without it we are left with implausible hypotheses for the advent of modern human symbolism and language: language capacities projected back a million years or more, recent magic-bullet language or symbol genes, or a purely cultural emergence of symbol use, with its indefensible hard border between nature and nurture. More and more archaeological evidence uncovered in recent years points toward a gradual accrual of modern symbolic capacities, especially in early *Homo sapiens* but likely in our Neandertal cousins and perhaps in other hominin groups. By the time the ancestors of today's humans began to expand through Africa and the rest of the world, some 60,000–90,000 years ago, it is probable that they exploited fully modern symbolic capacities.

Hyperindexicality suggests a gradation of semiosis among birds also. It seems unlikely that the semiotic complexity of birdsong (and also of birds' toolmaking, in particular among corvids) is equaled in nonsongish behaviors of suboscine passerines — passerines other than songbirds. Yet the presence in these, in comparatively undeveloped forms, of some of the brain structures hypertrophied in the song system of songbirds points to a neural architecture supporting some degree of semiotic capacity. Among mammals, an analysis of vocal-learning animals — elephants, cetaceans, seals, and a few others in addition to humans — would probably reveal degrees of semiotic complexity, with *Homo sapiens* a case of particular hypertrophy. Differences among songs and click codas in whales and dolphins might also suggest grades of semiotic complexity, though here, as with cephalopods, knowledge is so hard to come by that even the mode of production of their "vocalizations" is for many species not fully understood.

TWENTY-SIX

Questions Concerning

Technology

The distinction of a realm of semiotics and meaning within the causal information that pervades the biosphere can shed light on the nature of tools and technology. Toolmaking was once thought to be a marker of human achievement, even the decisive innovation that launched the hominin clade on its march toward modernity some three million years ago. We now understand that toolmaking is widespread among nonhuman animals (Shumaker, Walkup, and Beck 2011). Among primates, capuchin monkeys break nuts open using a hammerstone and stone "anvil," orangutans hold leaves to their lips to alter acoustically their calls, and chimpanzees probe tree trunks and termite mounds with sticks to pull out insects. Among cetaceans, bottlenose dolphins don sponges on their beaks to protect them as they forage for bottom-dwelling fish (Smolker et al. 1997). Among birds, corvids are the technological wizards, showing complex toolmaking aptitudes in experiments; in the wild, New Caledonian crows impale insects on sticks they have trimmed for the purpose (Chappell and Kacelnik 2002; Bluff et al. 2007). There are even reports of tool-wielding octopuses (Finn, Tregenza, and Norman 2009).

This dispersion of toolmaking well beyond humans presses on us several questions concerning the nature and extent of technology. First, toolmaking involves an exploitation of affordances in the niche. How are we to understand the difference between niche construction in general, a condition of all life framed in the second evolutionary abstract machine, and niche construction employing tools? Here we can sense the pressing danger of a slippage whereby a bird using a tree for a nest site, a parasitic wasp laying eggs in a caterpillar, a plant absorbing nutrients from the soil, even an amoeba engulfing another microbe would all be considered tool users. Yet tools seem to represent a special category of animal behaviors exploiting affordances, more limited than this sweeping view suggests. What features define this category?

At the same time as they delimit tool use, these features will need to expand it far beyond humans. But most modern characterizations of technology, reaching back in a line through Bernard Stiegler to Gilbert Simondon and then to Walter Benjamin and Martin Heidegger, have focused on toolmaking as a human propensity or capacity. How can we overcome anthropocentrism to achieve a general, transspecies understanding of technology?

In Section 7 we saw how the affordance horizons of organisms can be altered by the reciprocal interactions between homeostatic regulation within organisms and the changing niche outside them. Tool use seems to shift the horizon in a way subtly different from this. It forces the issue, remaking affordance possibilities through an *intentful* manipulation of materials, where that adjective stands in, for the moment, as a placeholder for other words. What capacities enable this?

Already in the 1920s Heidegger conceived the use of tools — or *equipment*, in the standard translation of his *Zeug* — in terms anticipating the affordances of niche-construction theory: "In the environment certain entities become accessible which are always ready-to-hand" (Heidegger 1962, p. 100). This "readiness-to-hand" (*Zuhandenheit*) is how a tool manifests itself through the enactment or performance of the work it does. The performance shapes the tool user's "concernful dealings" with the environment in such a way that some material is used for something (pp. 96, 100), and thus the tool becomes an encounter of the Being of its user with the world.

It engages the user as an aspect of *Dasein*, Being-in-the-world, in Heidegger's famous term. In this engagement tools can be seen to be processual: they become tools only as they cross with Being-in-the-world as parts of a worldly network, a *net* of circumstances bringing about whatever *work* their readiness-to-hand facilitates. Every instance of readiness-to-hand is a small portal onto the totality of Being, and tools are experienced as affordance processes in a larger niche, never in their discreteness as hammer, stick for foraging, and so forth.

In these processes Heidegger recognized a specificity of tools in relation to what they do, an "in-order-to" or "toward-which" of any tool that manifests its *reference* or *assignment (Verweisung)* to a particular kind of work (pp. 97–99). The reference of a tool is a pointing of it toward the work that manifests it as tool: the crow's stick points to the grub-impaling it enables, the orangutan's leaf to the altered vocalizations produced, and a hominin's Acheulean hand ax or kitchen knife to the cutting it achieves. This means that tools are Peircean indexes indicating the jobs they do. Readiness-to-hand — in the expanded view, readiness-to-hand, -beak, -paw, perhaps even -tentacle — is a sign, and the encounter with it constitutes a tool via a semiotic process. Tools are irreducibly semiotic.

To say that tools are semiotic processes, Paul Kockelman has written in an analysis joining Heidegger to Peirce, "is to say that they consist of signs, objects, and interpretants" (2015, p. 178). In Heidegger's discussion of the reference that lets "what is ready-to-hand be encountered," we see the outlines of a Peircean semiosis, one that limns both of its basic ingredients, the metarelation and the interpretant. "A sign is not a Thing which stands to another Thing in the relationship of indicating," Heidegger writes, seeing beyond simple relationality; instead it *"raises a totality of equipment into our circumspection so that together with it the worldly character of the ready-to-hand announces itself"* (Heidegger 1962, p. 110). The metarelation of semiosis appears here as the perception (circumspection) of any specific readiness-to-hand in relation to the totality of all equipment (or affordances-as-tools); this contextualized percept enables its specific, recursive reference to appear to the tool user. This creation of the tool-sign involves the shape of a niche, requiring "that it be possible for one's particular environment to announce itself for circumspection at any time by means of something ready-to-hand." The sign must have this niche affordance conducing to the revelation of readiness-to-hand: "Our circumspective dealings in the environment require some equipment ready-to-hand which in its character as equipment takes over the 'work' of *letting* something ready-to-hand *become conspicuous*" (p. III).

Readiness-to-hand thus comes to the tool user as an announcement shaped by the niche. In this we can discern one side of the Peircean interpretant: the address of a sign to the perceiver. This takes the form of a direct calling by the sign vehicle and, mediated through it, an indirect calling by the object, but it does not complete the semiotic process. Only in the perceiver's (tools user's) response, Heidegger's "circumspective dealing" with the world, is the tool created and the tool-sign consummated. Together these two vectors complete the full call-and-response dynamic of the interpretant. Tools are indexes in the form of material prostheses, but they are not given as such, even in their readiness-to-hand. Instead, they rise above the affordance horizons of the animals that use them through the reciprocal callings that form their interpretants, joining the material of the tool (sign vehicle) to the work of the tool (object) in the percept of an accessibility and opportunity (interpretant).

Understanding that tools are constituted through a semiotic process lodges the question of technology within the realm of meaning. This habitation for tool use can readily extend as far as we need to go to encompass nonhuman tools in all their variety. It can extend, that is, to the border region between tool use and the nonsemiotic, causal affordance relations basic to niche construction. Technology takes shape as semiosis, and tools, as indexes, participate in the specific kind of semiotic experience that extends beyond humans. (Nonhuman animals, we remember, likely do not form the abstractions necessary to make icons, and they approach the rule-governance of symbols only rarely, in hyperindexical arrays such as those found in birdsong.)

Without the recursion and metarelation of the sign, without the special percepts formed in the call-and-response of readiness-to-hand, organisms' material relations with their environments exploit affordances not as tools but along the causal-informational pathways ubiquitous in niche construction. Judging close cases at the border between the semiotic and causal exploitation of material affordance can be as difficult as tracking the limits of semiosis itself, but, as with semiosis in general, the difficulty does not lessen the importance of the difference. A crow's sharpening a twig to impale grubs is tool use; the crow forms an interpretant relating the twig to the work it will enable. Most birds' nest making, on the other hand, seems to be a programmed, innate set of responses without signs or content. But what of the case of the mature male satin bowerbird, which not only builds an especially elaborate nest to attract a mate but also decorates it with a collection of found objects, all in a similar shade of blue (Prum 2017)? Here, perhaps, we have crossed over into the semiotic realm and, if so, the signifying trinkets might be thought of as tool-signs. But we need to move cautiously toward such an interpretation, since the elaborations of innate mating displays among birds are immensely varied and complex. Where in the bowerbird's selection of blue bits and pieces is there evidence of interpretant formation and of the signifying metarelation?

Discerning the semiotic essence of tools breaks down the brand of humanist parochialism that imagines technology to be a purely human propensity, and in the same motion it poses an extrahuman limit for tool use — the limit at the far border of sign making among animals. This in turn reveals the misstep of certain recent philosophical moves that explode Heidegger's readiness-to-hand into an undifferentiated relationality between any two objects across the whole cosmos (Harman 2002, 2011). Such "object-oriented ontology" aims to retain the referentiality Heidegger saw in tools, but in an immensely expanded form superseding Dasein: a black hole *refers to* a star it draws into it, as the star likewise refers to it; a paper screen refers to the dust that falls on it or the fire that consumes it as two among any number of "for-the-sake-of" relations in its world of inter-actions with other objects. By virtue of this supposed reference, these relations perform *as meaning* the "tool-being" of any object: "For the tool, to be is to *mean*" (Harman 2002, p. 25). We see here a flattening of the metarelationality of sign and meaning into a simple relationality, something Heidegger himself warned against. There is no place for the interpretant, and indeed meaning in tool-being is said to require no awareness or percept at all, since "no humans need exist in order for the paper screen to resist dust or perish by fire" (p. 34). Such meaning draws near to sci-fi panpsychism in its particular brand of semantic universalism, far exceeding even the teleosemantics and teleodynamics reviewed in Part II (see Tomlinson 2016). But it is clear that the relations of objects said to be meaningful by object oriented ontologists mostly entail causal information, not meaning at all.

We need to be cautious, also, in applying the idea of technol-ogy in extended, metaphorical ways. Writing, in the conventional sense of the word, certainly qualifies as a technology or many tech-nologies, constructed from material indexes pointing beyond it to ideas, things, or spoken language itself, and it was once customarily achieved using tools that were themselves clearly indexical: stylus and wax or clay, pen or repurposed bird feather and paper, and so forth. Yet the Derridean extension of writing (Derrida 1976) to encom-pass something like all the marks of difference that occur in the world — put there not only by humans or even only by living things, and certainly extending beyond any notion of intention — carries the concept into a different category where technology no longer applies or, rather, where all technology is embraced, along with semiosis, meaning, and ultimately causal information as well. Ignoring this category shift conflates Derrida's arche-writing with the far nar-rower categories of tool, technology, sign, and meaning created by the niche construction of certain animals, opening a path once more toward an undifferentiated semantic universalism. Arche-writing, unlike conventional writing, is not necessarily semiotic, meaningful, or tool-like at all. It is divided into two realms, like information: a purely causal arena of difference and within it a semiotic one.

A knife or foraging stick may point clearly enough to the work they do, but such clarity seems not to obtain in all tools. In what fashion does an internal combustion engine point? A nuclear power plant? Why, generally, is the indexical nature of machines so frequently hidden from us? We might think that the difficulty arises from the systematic interrelation of parts in complex *technical objects*, analyzed in the 1950s by Gilbert Simondon (2017). This could suggest that technics sponsors an increase in complexity that obscures machines' semiotic foundation, diverting us from their indexicality toward their systematic organization. It is true that not all composite or systematic tools lose in this way a clear sense of indexical pointing. A Paleolithic spear, composed of wooden shaft, stone point, leather or grass straps, and adhesive to fix the point on the shaft, does exactly the opposite, clarifying the indexicality of the tool in a way that no one of its components can do individually. Nevertheless, even technical objects simpler than an engine or power plant can forfeit their indexical clarity. A triode, to take one of Simondon's favorite examples, does not seem to point to the signal amplification it makes possible; neither do the anode, cathode, intervening screen, electrical circuit, filament, or vacuum tube that it comprises.

What seems to be at work here is a replacement of clear indexicality with an emergent function that arises in part from what Simondon calls the *milieu associé*, the context associated with the tool. This is equivalent neither to Heidegger's world nor to my niche but is instead a two-sided thing, partly environmental conditions in which a technical object operates, partly the distinctive human creativity or inventiveness that Simondon features in his theory. And we might glimpse more clearly still the remnant of primordial pointing in another category Simondon introduces, the abstract *technical essence* of even complex machines, concretized in their material realization. The essence of the triode is the "asymmetrical conductance"

(Simondon 2017, p. 45) of electricity found already in the simpler diode — anode and cathode without intervening screen — and this asymmetry points toward the fundamental inequality of amplification, perhaps as clearly as the length and pointedness of a spear point toward its piercing function. The concretization of the triode as amplifier, on the other hand, dependent on the control of the asymmetry enabled by the screen as well as on the other parts of the device, distances us from its technical essence and its indexical pointing.

Perhaps, alternatively, we should think of complex technical objects as internalizations of indexicality, lodging pointing functions in the interrelations of individual components. In this view a cog in an automotive transmission points to other cogs with which it meshes, but the indexicality remains fixed at this level. This suggests an analogy between such objects and birdsong — or even human musicking, in which the meaning of the whole comes about from the play of expectation and fulfillment that arises from the pointing of individual gestures to one another, usually on several hierarchic levels. The gestures can involve scalar pitch, melodic shape, harmonic entities, rhythmic patterns usually within larger metrical frames, and timbral variety. The machine/music analogy is certainly not a perfect one, first because no machine is cognized by the human brain in the moment-to-moment flux of its components the way music is. Nevertheless, in musicking we can see the enactment of indexicality at several levels — music is arguably the most highly developed hyperindexicality universal to modern humans — and something rudimentarily like this might characterize complex machines and explain the obscuring of simple tool-pointing in them.

The analogy suggests also that complex tools, even the Paleolithic spear, can be thought of as hyperindexical material arrays that do not easily transfer the indexical functions of their components onto meanings at any higher hierarchic level of the whole machine. What we might call the *technical paradox*, then, consists in a composite hierarchization that obscures the semanticity of its parts. Complex machines steer a path parallel to that along which semiosis results in the most complex signs, hyperindexical arrays and symbols, but they do so in a way that distances machinic function from its roots in semiosis. Perhaps, in fact, this paradox is the root of our difficulty in defining just what a tool is.

What, finally, founds the toolmaking abilities of some animals? To ask this is to ask what is connoted by such Heideggerian words and phrases as concernful dealings, circumspection, and circumspective dealings, all of them seeming placeholders for qualities of Being, as well as to ask also what, in my own placeholder, is meant by intentful manipulation. The indexical nature of tools leads to the answer this book has urged across a broader terrain. Toolmaking exploits material affordances in a variety of sign-making, and so is dependent on the neural systems and resulting capacities that mark off semiotic animals from all other organisms: the reciprocal networks of processing in neural systems divided into multiple levels and nuclei, and, arising from them, episodic memory, situational learning, and complex forms of attention. Only animals showing such capacities make tools, and they do so under the aegis of the meanings they create in their manifold beings-in-the-world.

TWENTY-SEVEN

Questions Concerning

Culture

Over the last thirty years or so, the existence of culture in animals other than *Homo sapiens* has been proposed repeatedly and debated heatedly (for examples, Galef 1992; Wrangham et al. 1994; Rendell and Whitehead 2001; Laland and Hoppitt 2003; Laland and Galef 2009; Whitehead and Rendell 2014). The discussion naturally involves detailed consideration of the behaviors of the animals in question, but it hinges also on the definition of culture itself. This varies widely across a spectrum from anthropocentric definitions that effectively limit culture to humans to broader definitions, postanthropological or posthuman, that embrace many species. Posthuman conceptions aim to encompass the common behavioral ground among animals such as apes, elephants, cetaceans, and songbirds that seems to reflect capacities once thought to be uniquely human (Rendell and Whitehead 2001). A secondary aim is to illuminate the emergence of culture among our hominin ancestors, which we are ill equipped to do with any narrow conception of culture (for different approaches to this question, see Richerson and Boyd 2005; Chase 2006; Laland 2017; Tomlinson 2018).

The conception of meaning in this book inserts semiosis into the discussion and suggests a hierarchic relation between signs and culture. Here the sphere of culture is nested within the sphere of semiosis as a particular way of marshaling and exploiting signs; the sphere of symbolism, a particular type of semiosis, is further nested within the sphere of culture. All of these are nested within the far larger sphere of animal sociality (see Figure 27.1). Stated another way, most sociality in the animal world is nonsemiotic, most semiosis is noncultural, and most culture is nonsymbolic, therefore indexical. We need to unpack these nested realms.

The anthropological concept of culture that emerged in the eighteenth century and held sway through the twentieth was built on two primary components: the content or *stuff* of a culture and the *processes* of transmission by which it is conveyed and stabilized across generations (Williams 1976). We can speak of an *archive* of stuff in a given culture, on any scale from local to global, and the ingredients of such archives named in anthropological discussion are usually various and inclusive: patterns and systems of behavior, practices, works, meanings, and symbols — these are some of the most frequently named

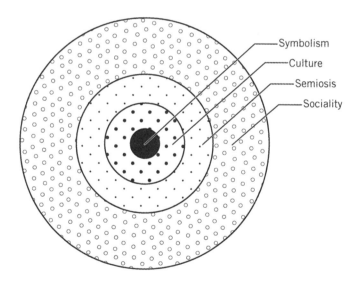

Figure 27.1. Nested spheres of animal communication.

categories — as well as gestures, techniques, attitudes, values, rituals, ideas, conceptions, theories, and more. All these mark the semantic nature of culture, its essential aboutness. This is affirmed in the process of transmission built into the anthropological concept, which involves some kind of enactment and imitation, that is, teaching and learning directed to and receptive of it. Such processes impart and acquire not causal but semantic information, not relation itself but metarelational percepts, even when they take the form of material practices like toolmaking imitated and reproduced. Culture is culture *about*, just as signs are signs *of*, and the transmission of culture conveys meaningful stuff. Like tools but on a broader scale, culture is semiotic.

The anthropological culture concept is axiomatically anthropocentric, as any definition of it in the literature reveals (for a compendium of examples see Baldwin et al. 2006; for two classic accounts, see Kroeber and Kluckhohn 1952 and Geertz 1973). Nevertheless, it has bequeathed its two primary components, stuff and process, to the new, posthuman culture concept. In order to identify a nonhuman culture we need to discern both its mechanisms of transmission and its transmitted archive. This requires that we identify its semiosis, the interpretant formation incorporating analysis of aspects of sign vehicle and object that come to relate them in perception. Semiosis arises in animal sociality wherever metarelational signs augment innate, causal signaling — where, for example, a learned birdsong tells another bird of mated status, territorial claim, relative strength or aggressiveness, and so forth, but not where a pheromone emitted by a waggling worker bee sets off a chemical cascade inducing another bee to forage. The honeybee example involves complex sociality without semiosis and inhabits the largest sphere of Figure 27.1, while the songbird example lodges in the next smaller sphere.

The question of nonhuman culture concerns the move to the third sphere, marking the difference between a noncultural semiotic animal and a cultural one. Culture does not follow automatically from semiosis but requires further processes that bring characteristic consequences. Here one aspect of the legacy of the anthropological culture concept comes to the fore: the relative stability of the transgenerational archive. This is not a frozen stability, of course; cultural archives shift across generations, and culture, human or not, is always a body of aboutness durably recognizable but in motion, a *changing same* (Gilroy 1991). Nevertheless, the relative stability brings with it three features characteristic of cultural archives (Tomlinson 2018). They involve, first, some degree of *abstraction* of the signs in them from their immediate deployment and use, a displacement of a sign away from in-the-moment social transaction toward what Maurice Bloch (2013) calls the "transcendental social." Once abstraction sets in, it fosters another tendency of cultural archives: the *accumulation* of distinct signs. In nonhuman cultures this attains only a shallow depth, at least in comparison to human culture, but shallow or deep it is always important because of the juxtapositions in it of abstracted signs. The distinctiveness of these — the differences marked among signs in any abstracted array — conduce to a third feature, *systematization*. This tendency is marked even in a minimal system of difference: One gesture indicates that you may approach with impunity, a distinct one suggests caution; one birdsong or even one syllable in a song indicates a readiness for sex, while a different one indicates mutual readiness on the part of a potential mate. Systematization in an array of signs points along the path leading to syntactic arrangement among them, but unless we widen the idea of syntax to embrace any interaction of contrastive meanings it is not a requirement of cultural archives. Even in the absence of syntax, we approach in the abstracted, accumulated, and systematized stuff the semiotic state of hyperindexicality, and it may be that this names a minimal requirement of every archive of signs, hence of culture itself.

Why did the semiotic content of culture not suggest, within the anthropological concept, an extension beyond the human? Clifford Geertz offers an exemplary case in point, one that spotlights the next differentiation in our nested spheres, between the cultural and the symbolic. For Geertz culture was a "historically transmitted pattern of meanings embodied in symbols, a system of inherited conceptions expressed in symbolic forms," a culture concept he considered to be "essentially semiotic." "Believing . . . that man is an animal suspended in webs of significance he himself has spun," he wrote, "I take culture to be those webs" (Geertz 1973, pp. 89, 5). We wonder here at two gestures, the quick slippage from animal to "man" — fifty years on it can come as a shock to see that earlier humanisms regularly denied *any* signification to nonhuman animals — and the restriction of culture to symbols. The two are logically related. Geertz spirals quickly in on human exceptionalism because his semiotics concerns symbols alone, with little accommodation for indexes. His culture concept makes a brief nod to semiotic theory and to Peirce, but it is more deeply rooted in Ernst Cassirer's symbolic forms and Susanne Langer's aesthetic symbolism, both ardently anthropocentric. This celebration of human symbolism was not likely to sponsor an expansive culture concept opening a continuous line from human to nonhuman, and it left anthropologists' human animal isolated on a symbolic island. Human exceptionalism remained a powerful theme of late twentieth-century human science, recurring even in accounts more attentive to both semiotic and evolutionary theory than Geertz's, for example Terrence Deacon's important book of 1997 *The Symbolic Species: The Co-evolution of Language and the Brain.*

If instead we think of cultural signs as more than exclusively symbolic, discerning then indexical cultures, we can open a wide avenue for understanding nonhuman culture. This reveals a sphere of culture larger than the sphere of symbolism and grants a vantage on human culture itself from which it is seen to be as dependent on indexes as it is on symbols. This has immediate consequences for our understanding of the long history of hominin cultural capacities. We can point, for example, to two universal modes of expression in modern human culture in which hyperindexical ordering is foregrounded: musicking and ritual. The ancient roots of each of these reach to our presymbolic, presapient past, where their indexicality grew complex alongside likewise indexical modes of protolanguage and protodiscourse (Tomlinson 2015, 2018). These are features of deep human history made difficult of access by any human science that starts with symbols as the elementary particles of culture. (Of the role of icons in human culture there is also no question, though the abstract conceptualization involved in true iconicity, described in Section 12, probably requires a symbolic basis not required for indexicality, making iconicity a relatively recent addition to hominin semiosis.)

The posthuman culture concept, arching across species and deep human history in a way its anthropological counterpart cannot, first took shape in the 1980s, not in the human sciences but in a new field attempting to understand the interactions of genomes and cultural practices. Gene-culture coevolution began as a dual-inheritance theory, an effort to analogize cultural transmission and genetic inheritance and to model the first using the quantitative tools developed for the second in midcentury evolution studies (see Boyd and Richerson 1985; Cavalli-Sforza and Feldman 1981; for a review, Tomlinson 2018). At first the analogy tended to maintain some distance between genes and culture, but early on it became clear that this could not hold — that interrelations between the two kinds of inheritance needed to be charted (Durham 1991). In the 1990s the relations tended to be thought of in lopsided terms, with genetic inheritance mainly shaping cultural inheritance. The coalescing of niche-construction theory, however, with its mechanisms of feedback from the inherited niche to the genome, required a differently balanced view in which cultural inheritance, part of a cultural animal's niche, could alter genomes just as they could alter it (Odling-Smee, Laland, and Feldman 2003). This revealed the full "coevolutionary dance" between genes and culture, engaged wherever "cultural traditions create novel environments . . . that can affect the fitness of alternative *genetically* transmitted variants" (Richerson and Boyd 2005, p. 190).

Gene-culture study settled in the 1980s on a very broad definition of culture as "the transmission from one generation to the next, via teaching and imitation, of knowledge, values, and other factors that influence behavior" (Boyd and Richerson 1985, p. 2). Evident here are the two features inherited from the anthropological concept, the stuff and the process, which together bring about the intergenerational persistence of cultural aboutness: the archive. Among partisans of nonhuman culture, this definition has been echoed, cited, and adapted widely enough for it to appear as a consensus. Here are several examples, quoted from a compendium offered by Luke Rendel and Hal Whitehead (2001, p. 310). Culture is behavior "that is transmitted repeatedly through social or observational learning to become a population-level characteristic"; "behavior patterns shared by members of a community that rely on socially learned and transmitted information"; "behavior or information with two primary attributes: it is socially learned and it is shared within a social community"; "shared behavior or information within a community acquired through some form of social learning from conspecifics"; or "behavior that is (a) transmitted socially rather than genetically, (b) shared by many members within a group, (c) persistent over generations, and (d) not simply the result of adaptation to different local conditions." The definition having been settled, the discussion usually turns to the nature and mechanisms of social learning (that is, process) and descriptions of learned behaviors shared in specific animal communities (stuff).

In several of these definitions the transgenerational persistence of cultural stuff initially specified by Richerson and Boyd falls away, leaving social learning and sharing within a community as the criteria for animal cultures. These are important criteria, but focus on them alone can lead to an underdetermined and overly broad culture concept in which all semiosis among members of a community amounts to culture, in effect conflating the semiotic and cultural spheres of Figure 27.1. It diverts attention away from longitudinal or historical examination of the archives formed in the interaction of process and stuff, and in fact most discussions of the posthuman culture concept omit general consideration of the lineaments of these archives, whether conceived as the outcome of abstraction, accumulation, and systematization or along other lines (for an exception, see Tomasello 1994).

These archival dynamics are, however, decisive for a posthuman concept if it is to distinguish sociality conducive to culture from the capacious realm of sociality in general. For this reason, the posthuman culture concept requires another stipulation added to Boyd and Richerson's definition: Culture is the transmission from one generation to the next, via teaching and imitation, of an *accumulated, systematized archive* of knowledge, values, and other factors that influence behavior. And it needs also, finally, a rewording that can generalize beyond anthropocentric categories such as knowledge and value: Culture is the transmission from one generation to the next, via teaching and imitation, of an accumulated, systematized archive of *signs* that influence behavior.

In the context of this adjusted definition, we can survey the range of animal lifeways extending from causal information without sociality to culture in five exemplary episodes. Many animals live their lives through causal information alone. They are of course entangled with their niches in rich and complex ways, like all life-forms, but they feed, grow, and reproduce without social interactions with conspecifics. Such *asociality* stands entirely outside the spheres of Figure 27.1. It characterizes sea anemones, for instance, opportunistic, mostly sessile predators reproducing either asexually or sexually without social interaction. The whole phylum to which they belong, the cnidarians, probably inhabits this category, alongside other phyla such as porifera (sponges) and echinoderms (starfish, sea urchins, and the like) — though in this last case some species show an incipient sociality that might move them to the next category.

A vast number of animals show *asemiotic sociality*, some degree of social interaction with conspecifics without semiosis; these inhabit the largest sphere in Figure 27.1. The case of honeybees shows how elaborate such sociality can become. It is a special instance among more than a million others from the phylum Arthropoda alone that likely belong in this category. Whole other phyla probably belong here too: annelids or segmented worms, and perhaps all mollusks — or most of them, leaving room for the possible exception of cephalopods. Among vertebrates, we have seen, semiosis might not extend far beyond birds and mammals. If so, many reptiles, amphibians, and fish would also fall into this category. There may even be nonsemiotic mammals and birds that belong here.

The heart of the question of culture, I have suggested, comes in the distinction between semiosis without culture and with it, between *acultural* and *cultural semiosis*. Among mammals and birds, baboons offer an instance of acultural semiosis that is significant because of their especially complex sociality (Cheney and Seyfarth 2008). In baboon lives, signs form a constant medium, deployed in the interactions within the troop to negotiate hierarchies of matrilineal status, to rank-order access to food, to protect the troop from predators and babies from nonpaternal adult males, and more. These vocal and bodily gestures include welcoming signs, warning signs, and aggressive and submissive signs. They are to some degree learned from experiences in the troop, thus fulfilling the social learning aspect of the posthuman culture concept, but they occur only in the here-and-now dimension of Bloch's social transaction. There is little evidence that they are abstracted to form a body of evolving signs passed and developed from generation to generation, hence little evidence of the accumulation of a cultural archive. These semiotically adroit animals learn much from immediate interactions with one another and remember much for use in future interactions, but it seems that they do not form cultures.

Many nonhuman animals have by now been proposed as cultural animals — inhabitants, then, of the sphere of cultural semiosis. These include many songbirds; a number of different cetaceans, among which the best-studied are bottlenose dolphins, killer whales, sperm whales, and humpback whales; and various monkeys and apes, with chimpanzees in the research lead here. But discerning the threshold to culture in the posthumanist sense has proved, for the symbolic species observing all this, a daunting task.

For these animals we have ample evidence of social learning and the resulting spread of learned behaviors across discrete populations. Both are evident in many kinds of songbirds, and the dialects their songs sometimes fall into within local groups were among the first nonhuman behaviors advanced as instances of culture (Trainer 1989). Social learning is likewise witnessed among humpback whales, whose songs have fascinated us since their discovery in the 1960s (Payne and McVay 1971). They resemble birdsongs, except exploded to a scale worthy of their makers (Whitehead and Rendell 2014; Garland and McGregor 2020): each song lasting as long as half an hour, joined in song bouts lasting as long as a day, heard and responded to across a breeding ground of a thousand square kilometers, and carrying across even longer distances. These songs are shared among the males of large populations numbering in the thousands, inhabiting whole ocean basins. They are learned, with the delicate precision of birdsong, as segmented, hierarchic, combinatorial structures of units, phrases, and themes that make up the songs, which in turn make up the bouts. All this is adopted by the males in the large group, performed and precisely repeated, but still the song is not unchanging. Innovative phrases and themes, sometimes imported by "minstrel" whales from other populations, are interpolated in the home structures (by which whale? why and with what authority? — we don't know), then adopted and spread through the whole group (Noad et al. 2000; Garland et al. 2017).

Birdsong dialects and humpback whale song types represent the semiotic machine at work — signs socially learned and shared in a community. They even appear to manifest something like traditions, though ascertaining this requires evidence of both persistence and shift across generations that has been hard to come by. The variants that distinguish a local birdsong dialect remain closely bound to the syllables of the song types native to the species in question. Whatever the innate constraints on these syllables, there is no evidence of an intergenerational cultural tradition that can accumulate and diverge from them. Neither is there evidence that the songs of humpback whales form coherent intergenerational archives — evidence almost impossible to collect, given the large, migratory populations and the long individual lifetimes involved. At any rate, the whales' songs change enough that those of a population are essentially unrelated to their predecessors within five years or so. Is this the stuff of cultural archives or not?

Chimpanzees provide the solidest evidence we have for nonhuman culture. They are animals whose capacity for social transmission can be put to experimental test, and they reveal a ready transmission of newly learned behaviors among a group (Whiten et al. 2007). Observation of chimps in the wild now has a long and rich history, and this has shown again and again that socially transmitted behaviors dispersed through one group can differ from those in another. One synthesis of fieldwork at seven different sites tabulated thirty-nine distinctive learned behaviors prevalent in one or more populations while absent from others, including habits of grooming and many varieties of tool use (Whiten et al. 1999). In each of these cases there were no obvious differences of environmental affordances from group to group that could explain the behavioral differences. The length of the studies of each group, totaling 151 years of observation, seems to ensure that we are in the presence of true transgenerational traditions of learned behaviors — cultural stuff or archives of some depth, in other words. And the record of chimpanzee cultural attainment continues to grow (Whiten 2017).

It seems unlikely that chimpanzees are alone among nonhuman animals in forming cultural archives. The unique clarity of their case reflects the diversity of their cultural behaviors, which contrasts with the shallower, less varied archives that have been proposed for other animals — those birdsong dialects and communal humpback whale songs, for instance. At the same time, the case of chimpanzees sets in relief the greater challenges involved in affirming culture in other animals, where less pronounced archives may be harder to track in longitudinal study because of the habitats and lifeways of the animals involved.

Arching over these pragmatic difficulties is the problem of defining culture in a way that can capture what is special about it as a set of animal attainments while not sweeping into its orbit aspects of animal life not specific to it. In anthropology, especially since the advent of critical and postcolonial approaches in the 1970s, similar difficulties in dealing with human cultures have seemed to some to justify abandoning the culture concept altogether. But doing so at the moment of its posthuman extension seems capricious and almost unjust, a bit like changing the rules of the game just when a new kid in the neighborhood manages to win. And it would in any case undo the profitable blurring of the human/nonhuman border hard won in the extension. The diffuseness of the culture concept in its anthropological version is a consequence of the many kinds of complex behavior it has always attempted to embrace, hence a product of the unparalleled depth and variety of human cultural archives. In one obvious way the posthuman culture concept simplifies matters, since diversity of cultural behavior will not be found in nonhuman animals in anything close to a human degree. This narrows the kinds of evidence we seek.

At the same time, in widening the range of animals involved in the search, the posthuman culture concept encounters the diverse lifeways of animals as far apart as birds and whales. Its own, special diffuseness is an expression, more direct than in the anthropological concept, of the entanglements across time and space of niche and animal. The changing same of a culture is finally nothing other than a microcosm of the beguiling, bewildering changing same of the lifeways of animals producing it.

Here for a final time semiosis helps clarify difficult issues. Any cultural microcosm will have emerged from the operation of the semiotic machine, and this is more tractable than a putative culture machine could be, since its single, essential output — referentiality — arises from a delimitable process — perceptual analysis, metarelation, interpretant formation — that is reflected in behavior and associable with neural substrates. This is not to suggest that mapping the precincts of semiosis through the animal world will be easy; we have assayed this project and witnessed the kind of granular study it requires. But its rewards will be considerable: a new estimation of meanings humans share with others in the world, and a deepened appreciation of complexity built from different, meaningless processes.

Acknowledgments

Like many books emerging now, this one was shaped by the pandemic during which it was written. Detailing how this awful, awesome episode exerted its formative force on this little project is not for an acknowledgments page, except to say that it enforced a solitude that was at once productive and alienating. I was very lucky that alienation was the worst that Covid-19 brought my way.

The solitude was not unrelenting. It was interrupted by the Zoom meetings that quickly became a way of life in academia as elsewhere, and I am grateful for many conversations with and reactions from scholars and students when I presented materials related to the book at such events. A consultation with three readers of the manuscript, Jürgen Renn, Manfred Laubichler, and Matthias Schemmel, stands out among these, as does another with my discerning and bold editor at Zone Books, Ramona Naddaff. The solitude was broken also by cautious visits with the inhabitants of my small bubble: Ornella Rossi and Francesco Casetti, whose conviviality and conversation always revived my energies and reset my compass (Francesco also offered valuable comments on parts of the manuscript); Laura Suciu and Denis Pelli, whose many antidotes for pandemic included bocce and Bela Tarr; and my friend and sometime co-conspirator Günter Wagner, whose conversation always brought rigor to my thinking and on whom none of my disciplinary overreach should be blamed. I put final touches on the manuscript as I returned to face-to-face interaction with my Yale students, which even behind masks is a warmly rewarding experience.

Visitors closer still to the center of my bubble were my now-grown children, to whom this book is dedicated; I'm only sorry that so many of our interactions were at an electronic distance, too few in person. Invigorating the bubble always was the wonderdog Blitz, his enthusiasm a beacon. And there with me also, always, was Juliet Fleming, whose marvelous mind and presence are marked in whatever is right about the book. It's beyond me to imagine having spent the pandemic years with any other constant companion.

Works Cited

Abe, Kentaro, and Dai Watanabe. 2011. "Songbirds Possess the Spontaneous Ability to Discriminate Syntactic Rules." *Nature Neuroscience* 14:1067–74.

Abou-Shaara, H. F. 2014. "The Foraging Behaviour of Honey Bees, *Apis mellifera*: A Review." *Veterinarni Medicina* 59:1–10.

Alaux, C., A. Maisonnasse, and Y. Le Conte. 2010. "Pheromones in a Superorganism: From Gene to Social Regulation." *Vitamins and Hormones* 83:401–23.

Allen, Timothy A., and Norbert J. Fortin. 2013. "The Evolution of Episodic Memory." *Proceedings of the National Academy of Science* 110:10379–86.

Amy, Mathieu, Pauline Salvin, Marc Naguib, and Gerard Leboucher. 2015. "Female Signalling to Male Song in the Domestic Canary, *Serinus canaria*." *Royal Society Open Science* 2:140196.

Arendt, Detlev, Jacob M. Musser, Clare V. H. Baker, Aviv Bergman, Connie Cepko, Douglas H. Erwin, Mihaela Pavlicev et al. 2016. "The Origin and Evolution of Cell Types." *Nature Reviews Genetics* 17:744–57.

Atkin, Albert. 2013. "Peirce's Theory of Signs." *The Stanford Encyclopedia of Philosophy*, ed. Edward N. Zalta. http://plato.stanford.edu/archives/sum2013/entries/peirce-semiotics/.

Avarguès-Weber, Aurore, Theo Mota, and Martin Giurfa. 2012. "New Vistas on Honey Bee Vision." *Apidologie* 43:244–68.

Avey, Marc T., Leslie S. Phillmore, and Scott A. MacDougall-Shackleton. 2005. "Immediate Early Gene Expression Following Exposure to Acoustic and Visual Components of Courtship in Zebra Finches." *Behavioural Brain Research* 165:247–53.

Awh, Edward, Artem V. Belopolsky, and Jan Theeuwes. 2012. "TopDown versus Bottom-Up Attentional Control: A Failed Theoretical Dichotomy." *Trends in Cognitive Science* 16: 437–43.

Baddeley, Alan. 2000. "The Episodic Buffer: A New Component of Working Memory?" *Trends in Cognitive Sciences* 4:417–23.

———. 2003. "Working Memory: Looking Back and Looking Forward." *Nature Reviews Neuroscience* 4:829–39.

Balari, Sergio, and Guillermo Lorenzo. 2013. *Computational Phenotypes: Towards an Evolutionary Developmental Biolinguistics.* Oxford: Oxford University Press.

Baldwin, John R., Sandra L. Faulkner, Michael L. Hecht, and Sheryl L. Lindsley, eds. 2006. *Redefining Culture: Perspectives across the Disciplines.* Mahwah, NJ: Lawrence Erlbaum.

Barron, Andrew, Marta Halina, and Colin Klein. 2020–23. "The Major Transitions in the Evolution of Cognition." Templeton World Research Project. https://www.templetonworldcharity.org/projects-database/major-transitions-evolution-cognition.

Barron, Andrew B., Kevin N. Gurney, Lianne F. S. Meah, Eleni Vasilaki, and James A. R. Marshall. 2015. "Decision-Making and Action Selection in Insects: Inspiration from Vertebrate-Based Theories." *Frontiers in Behavioral Neuroscience* 9:216–29.

Barth, Friedrich G. 1982. *Insects and Flowers: The Biology of a Partnership.* Princeton, NJ: Princeton University Press.

Beckers, Gabriël J. L., Johan J. Bolhuis, Kazuo Okanoya, and Robert C. Berwick. 2012. "Birdsong Neurolinguistics: Songbird Context-Free Grammar Claim Is Premature." *NeuroReport* 23:139–45.

Beecher, Michael D., and John M. Burt. 2004. "The Role of Social Interaction in Bird Song Learning." *Current Directions in Psychological Science* 13:224–28.

Beecher, Michael D., John M. Burt, Adrian O'Loghlen, Christopher N. Templeton, and S. Elizabeth Campbell. 2007. "Birdsong Learning in an Eavesdropping Context." *Animal Behaviour* 73:929–35.

Beecher, Michael D., Akçay Çaglar, and S. Elizabeth Campbell. 2020. "Birdsong Learning Is Mutually Beneficial for Tutee and Tutor in Song Sparrows." *Animal Behaviour* 166:281–88.

Beecher, Michael D., and S. Elizabeth Campbell. 2005. "The Role of Unshared Songs in Singing Interactions between Neighboring Song Sparrows." *Animal Behaviour* 70:1297–1304.

Beecher, Michael D., P. K. Stoddard, S. Elizabeth Campbell, and C. L. Horning. 1996. "Repertoire Matching between Neighboring Song Sparrows." *Animal Behaviour* 51:917–23.

Beekman, Madeleine, Rosalyn S. Gloag, Naïla Even, Wandee Wattanachaiyingchareon, and Benjamin P. Oldroyd. 2008. "Dance Precision of *Apis florea* — Clues to the

Evolution of the Honeybee Dance Language?" *Behavioral Ecology and Sociobiology* 62:1259–65.

Beekman, Madeleine, and Jie Bin Lew. 2008. "Foraging in Honeybees — When Does It Pay to Dance?" *Behavioral Ecology* 19:255–62.

Belzner, Sandra, Cornelia Voigt, Clive K. Catchpole, and Stefan Leitner. 2009. "Song Learning in Domesticated Canaries in a Restricted Acoustic Environment." *Proceedings of the Royal Society B* 276:2881–86.

Benard, Julie, Silke Stach, and Martin Giurfa. 2006. "Categorization of Visual Stimuli in the Honeybee *Apis mellifera*." *Animal Cognition* 9:257–70.

Bergman, Mats. n.d. *Fields of Signification: Explorations in Charles S. Peirce's Theory of Signs.* Philosophical Studies from the University of Helsinki 6. University of Helsinki: Department of Philosophy.

Bertalanffy, Ludwig von. 1969. *General System Theory: Foundations, Development, Applications.* New York: Braziller.

Berwick, Robert C., Gabriël J. L. Beckers, Kazuo Okanoya, and Johan J. Bolhuis. 2012. "A Bird's Eye View of Human Language Evolution." *Frontiers in Evolutionary Neuroscience* 4. https://doi.org/10.3389/fnevo.2012.00005.

Berwick, Robert C., Kazuo Okanoya, Gabriel J. L. Beckers, and Johan J. Bolhuis. 2011. "Songs to Syntax: The Linguistics of Birdsong." *Trends in Cognitive Sciences* 15: 113–21.

Bloch, Maurice. 2013. *In and Out of Each Other's Bodies: Theory of Mind, Evolution, Truth, and the Nature of the Social.* Boulder, CO: Paradigm.

Block, Ned, and Philip Kitcher. 2010. "Misunderstanding Darwin: Natural Selection's Secular Critics Get It Wrong." *Boston Review*, March 1.

Bluff, Lucas A., Alex A. S. Weir, Christian Rutz, Johanna H. Wimpenny, and Alex Kacelnik. 2007. "Tool Related Cognition in New Caledonian Crows." *Comparative Cognition and Behavior Reviews* 2:1–25.

Bolhuis, Johan J., Gabriel J. L. Beckers, Marinus A. C. Buybregts, Robert C. Berwick, and Martin B. H. Everaert. 2018. "Meaningful Syntactic Structure in Songbird Vocalizations?" *PLoS Biology* doi.org/10.1371/journal.pbio.2005157.

Bolhuis, Johan J., and Manfred Gahr. 2006. "Neural Mechanisms of Birdsong Memory." *Nature Reviews Neuroscience* 7:347–57.

Bortolotti, Laura, and Cecilia Costa. 2014. "Chemical Communication in the Honey Bee Society." In *Neurobiology of Chemical Communication*, ed. C. Mucignat-Caretta, chap. 5. Boca Raton FL: CRC Press/Taylor and Francis.

Bournonville, Catherine de, Aiden McGrath, and Luke Remage-Healey. 2020. "Testoster-one Synthesis in the Female Songbird Brain." *Hormones and Behavior* 121:104716.

Bowling, Daniel, and W. Tecumseh Fitch. 2015. "Do Animal Communication Systems Have Phonemes?" *Trends in Cognitive Sciences* 19:555–57.

Boyd, Robert, and Peter J. Richerson. 1985. *Culture and the Evolutionary Process*. Chicago: University of Chicago Press.

———. 2005. *The Origin and Evolution of Cultures*. New York: Oxford University Press.

Bradbury, Jack W., and Sandra L. Vehrencamp. 2011. *Principles of Animal Communication*. 2nd ed. New York: Sinauer Associates and Oxford University Press.

Brown, Culum, Martin P. Garwood, and Jane E. Williamson. 2012. "It Pays to Cheat: Tac-tical Deception in a Cephalopod Social Signalling System." *Biology Letters* 8:729–32.

Bueno, O., R.-L. Chen, and M. B. Fagan, eds. 2018. *Individuation, Process, and Scientific Prac-tice*. New York: Oxford University Press.

Campbell, Donald T. 1974. "Evolutionary Epistemology." In *The Philosophy of Karl Popper*, ed. P. A. Schilpp, 2 vols., 1:412–63. LaSalle, IL: Open Court Publishing Company.

———. 1983. "The General Algorithm for Adaptation in Learning, Evolution, and Percep-tion." *The Behavioral and Brain Sciences* 6:178–79.

Carroll, Sean B., J. K. Grenier, and S. D. Weatherbee. 2005. *From DNA to Diversity: Molecular Genetics and the Evolution of Animal Design*. 2nd ed. Malden, MA: Wiley-Blackwell.

Casetti, Francesco. 2015. *The Lumière Galaxy: 7 Key Words for the Cinema to Come*. New York: Columbia University Press.

Catchpole, C. K., and P. J. B. Slater. 2008. *Bird Song: Biological Themes and Variations*. 2nd ed. Cambridge: Cambridge University Press.

Cavalli-Sforza, Luigi Luca, and Marcus W. Feldman. 1981. *Cultural Transmission and Evolu-tion: A Quantitative Approach*. Princeton, NJ: Princeton University Press.

Chakraborty, Mukta, Solveig Walløe, Signe Nedergaard, Emma E. Fridel, Torben Dabel-steen, Bente Pakkenberg, Mads F. Bertelsen et al. 2015. "Core and Shell Song Systems Unique to the Parrot Brain." *PLoS One* 10, no. 6:e0118496.

Chappell, J., and A. Kacelnik. 2002. "Tool Selectivity in a Non-Mammal, the New Caledo-nian Crow (*Corvus moneduloides*)." *Animal Cognition* 5: 71–78.

Chase, Philip G. 2006. *The Emergence of Culture: The Evolution of a Uniquely Human Way of Life*. New York: Springer.

Cheney, Dorothy L., and Robert M. Seyfarth. 2008. *Baboon Metaphysics: The Evolution of a Social Mind*. Chicago: University of Chicago Press.

Cheung, Allen, Matthew Collett, Thomas S. Collett, Alex Dewar, Fred Dyer, Paul Graham, Michael Mangan et al. 2014. "Still No Convincing Evidence for Cognitive Map Use by Honeybees." *Proceedings of the National Academy of Science* 111:E4396-97.

Chittka, Lars. 2004. "Dances as Windows into Insect Perception." *PLoS Biology* 2:0898-0900.

Chittka, Lars, and Karl Geiger. 1995. "Can Honeybees Count Landmarks?" *Animal Behaviour* 49:159-64.

Chittka, Lars, and Jeremy Niven. 2009. "Are Bigger Brains Better?" *Current Biology* 19:R995-1008.

Clark, Andy. 2008. *Supersizing the Mind: Embodiment, Action, and Cognitive Extension.* Oxford: Oxford University Press.

Clayton, Nicola S., and Anthony Dickinson. 1998. "Episodic-Like Memory during Cache Recovery by Scrub Jays." *Nature* 395:272-74.

Collett, T. S., K. Fauria, K. Dale, and J. Baron. 1997. "Patterns and Places — A Study of Context Learning in Honeybees." *Journal of Comparative Physiology* A 181:343-53.

Creanza, Nicole, Laurel Fogarty, and Marcus W. Feldman. 2016. "Cultural Niche Construction of Repertoire Size and Learning Strategies in Songbirds." *Evolutionary Ecology* 30:285-305.

Cruse, Holk, and Rüdiger Wehner. 2011. "No Need for a Cognitive Map: Decentralized Memory for Insect Navigation." *PLoS Computational Biology* 7:e1002009.

Cummins, Robert, and Martin Roth. 2010. "Traits Have Not Evolved to Function the Way They Do Because of Past Advantage." In *Contemporary Debates in Philosophy of Biology*, eds. Francisco J. Ayala and Robert Arp, pp. 72-85. Oxford: Wiley-Blackwell.

Darmaillacq, Anne-Sophie, Ludovic Dickel, and Jennifer Mather, eds. 2014. *Cephalopod Cognition.* Cambridge: Cambridge University Press.

Darwin, Charles. 1888. *The Life and Letters of Charles Darwin, Including an Autobiographical Chapter*, ed. Francis Darwin. 3 vols. London: John Murray.

———. 2003. *The Origin of Species and The Voyage of the Beagle.* New York: Knopf.

Dawkins, Richard. 1982. *The Extended Phenotype.* Oxford: Oxford University Press.

Deacon, Terrence. 2012a. "Beyond the Symbolic Species." In *The Symbolic Species Evolved*, ed. Theresa Schilhab, Frederik Stjernfelt, and Terrence Deacon, 9-38. Berlin: Springer.

———. 2012b. *Incomplete Nature: How Mind Emerged from Matter.* New York: Norton.

———. 1997. *The Symbolic Species: The Co-Evolution of Language and the Brain.* New York: Norton.

Deamer, David W. 2020. *Origin of Life: What Everyone Needs to Know.* Oxford: Oxford University Press.

DeDeo, Simon. 2017. "Information Theory." Santa Fe Institute Complexity Intelligence Report 2. https://wiki.santafe.edu/images/9/9e/SFI_DeDeo_InformationTheory.pdf.

De Landa, Manuel. 1997. *A Thousand Years of Nonlinear History*. New York: Zone Books.

Deleuze, Gilles, and Félix Guattari. 1987. *A Thousand Plateaus: Capitalism and Schizophrenia*. Trans. Brian Massumi. Minneapolis: University of Minnesota Press.

De Marco, Rodrigo J., Mariana Gil, and Walter M. Farina. 2005. "Does an Increase in Reward Affect the Precision of the Encoding of Directional Information in the Honeybee Waggle Dance?" *Journal of Comparative Physiology* 191:413–19.

Dennett, Daniel C. 2017a. "A Difference that Makes a Difference: A Conversation." *Edge*. https://www.edge.org/conversation/daniel_c_dennett-a-difference-that-makes-a-difference.

———. 1995. *Darwin's Dangerous Idea: Evolution and the Meanings of Life*. New York: Simon & Schuster.

———. 2017b. *From Bacteria to Bach and Back: The Evolution of Minds*. New York: Norton.

———. 1983. "Intentional Systems in Cognitive Ethology: The 'Panglossian Paradigm' Defended." *Behavioral and Brain Sciences*. 6:343–90.

———. 1987. *The Intentional Stance*. Cambridge, MA: MIT Press.

Derrida, Jacques. 1976. *Of Grammatology*. Trans. Gayatri Chakravorty Spivak. Baltimore, MD: Johns Hopkins University Press.

———. 1978. *Writing and Difference*. Trans. Alan Bass. Chicago: University of Chicago Press.

Díaz, Paula C., Christoph Grüter, and Walter M. Farina. 2007. "Floral Scents Affect the Distribution of Hive Bees around Dancers." *Behavioral Ecology and Sociobiology* 61:1589–97.

Dickel, L., M. P. Chichery and R. Chichery. 1998. "Time Differences in the Emergence of Short- and Long-Term Memory during Post-embryonic Development in the Cuttlefish *Sepia*." *Behavioural Processes* 44:81–86.

Dornhaus, A., and L. Chittka. 1999. "Evolutionary Origins of Bee Dances." *Nature* 401:38.

———. 2004. "Why Do Honey Bees Dance?" *Behavioral Ecology and Sociobiology* 55:395–401.

Dretske, Fred I. 1988. *Explaining Behavior: Reasons in a World of Causes*. Cambridge, MA: MIT Press.

Dupré, John. 2012. *Processes of Life: Essays in the Philosophy of Biology*. Oxford: Oxford University Press.

Durham, William. 1991. *Coevolution: Genes, Culture, and Human Diversity*. Stanford, CA: Stanford University Press.

Dyer, F. C. 2002. "The Biology of the Dance Language." *Annual Review of Entomology* 47:917–49.

Dyer, Fred C., and Jeffrey A. Dickinson. 1996. "Sun-Compass Learning in Insects: Representation in a Simple Mind." *Current Directions in Psychological Science* 5:67–72.

Eco, Umberto. 1976. *A Theory of Semiotics*. Bloomington: Indiana University Press.

Eldredge, Niles, Telmo Pievani, Emanuele Serrelli, and Ilya Tëmkin. 2016. *Evolutionary Theory: A Hierarchical Perspective*. Chicago: University of Chicago Press.

Engesser, S., J. M. S. Crane, J. L. Savage, A. F. Russell, and S. W. Townsend. 2015. "Experimental Evidence for Phonemic Contrasts in a Nonhuman Vocal System." *PLoS Biology* 13: e1002171.

Esch, H., and J. Burns. 1996. "Dance Estimation by Foraging Honeybees." *Journal of Experimental Biology* 199:155–62.

———. 1995. "Honeybees Use Optic Flow to Measure the Distance of a Food Source." *Naturwissenschaften* 82:38–40.

Evangelista, C., P. Kraft, M. Dacke, T. Labhart, and M. V. Srinivasan. 2014. "Honeybee Navigation: Critically Examining the Role of the Polarization Compass." *Philosophical Transactions of the Royal Society B*. http.doi.org/10.1098/rstb.2013.0037.

Farina, Walter M., Christoph Grüter, and Paula C. Díaz. 2005. "Social Learning of Floral Odors inside the Honeybee Hive." *Proceedings of the Royal Society B* 272:1923–28.

Finn, Julian K., Tom Tregenza, and Mark D. Norman. 2009. "Defensive Tool Use in a Coconut-Carrying Octopus." *Current Biology* 19:R1069–70.

Fishbein, Adam R., William J. Idsardi, Gregory F. Ball, and Robert J. Dooling. 2019. "Sound Sequences in Birdsong: How Much Do Birds Really Care?" *Philosophical Transactions of the Royal Society B* 375:20190044.

Fodor, Jerry. 1990. *A Theory of Content and Other Essays*. Cambridge MA: MIT Press.

Fodor, Jerry, and Massimo Piattelli-Palmarini. 2010. *What Darwin Got Wrong*. New York: Picador.

Friday, Steven L., and Peter W. Greig-Smith. 1994. "The Effects of Social Learning on the Food Choice of the House Sparrow (*Passer domesticus*)." *Behaviour* 128:281–300.

Gahr, Manfred. 2000. "Neural Song Control System of Hummingbirds: Comparison to Swifts, Vocal Learning (Songbirds) and Nonlearning (Suboscines) Passerines, and Vocal Learning (Budgerigars) and Nonlearning (Dove, Owl, Gull, Quail, Chicken) Nonpasserines." *Journal of Comparative Neurology* 426:182–96.

Galef, Bennett G. 1992. "The Question of Animal Culture." *Human Nature* 3:157–78.

Garland, Ellen C., and Peter K. McGregor. 2020. "Cultural Transmission, Evolution, and Revolution in Vocal Displays: Insights from Bird and Whale Song." *Frontiers in Psychology* 11:544929.

Garland, Ellen C., Luke Rendell, Luca Lamoni, M. Michael Poole, and Michael J. Noad. 2017. "Song Hybridization Events during Revolutionary Song Change Provide Insights into Cultural Transmission in Humpback Whales." *Proceedings of the National Academy of Science* 114:7822–29.

Gavrilets, Sergey. 2004. *Fitness Landscapes and the Origin of Species.* Princeton, NJ: Princeton University Press.

Geertz, Clifford. 1973. *The Interpretation of Cultures.* New York: Basic Books.

Gibson, James J. 1979. "The Theory of Affordances." *The Ecological Approach to Visual Perception*, 119–37. Boulder, CO: Taylor & Francis.

Gil, Mariana, and Rodrigo J. De Marco. 2005. "Olfactory Learning by Means of Trophallaxis in *Apis mellifera*." *Journal of Experimental Biology* 208:671–80.

Gilroy, Paul. 1991. "Sounds Authentic: Black Music, Ethnicity, and the Challenge of a 'Changing Same'." *Black Music Research Journal* 11:111–36.

Giurfa, Martin. 2007. "Behavioural and Neural Analysis of Associative Learning in the Honeybee: A Taste from the Magic Well." *Journal of Comparative Physiology A* 193:801–24.

———. 2013. "Cognition with Few Neurons: Higher-Order Learning in Insects." *Trends in Neurosciences* 36:285–94.

Giurfa, Martin, Jeffrey A. Riffell, and Lars Chittka, eds. 2020. *The Mechanisms of Insect Cognition.* Lausanne: Frontiers Media SA.

Godfrey-Smith, Peter. 2009. *Darwinian Populations and Natural Selection.* Oxford: Oxford University Press.

———. 1994. "A Modern History Theory of Functions." *Noûs* 28:344–36.

———. 2017. *Other Minds: The Octopus, the Sea, and the Deep Origins of Consciousness.* New York: Farrar, Straus, and Giroux.

———. 2014. *Philosophy of Biology.* Princeton, NJ: Princeton University Press.

Gould, Stephen Jay. 1990. *Wonderful Life: The Burgess Shale and the Nature of History.* New York: Norton.

Gould, Stephen Jay, and Richard Lewontin. 1979. "The Spandrels of San Marco and the Panglossian Paradigm: A Critique of the Adaptationist Programme." *Proceedings of the Royal Society B* 205:581–98.

Graham, P. 2010. "Insect Navigation." In *Encyclopedia of Animal Behavior*, 167–75. Amsterdam: Elsevier.

Griffiths, Paul. 1993. "Functional Analysis and Proper Functions." *British Journal for the Philosophy of Science* 44:409–22.

Grusin, Richard. 2015. "Radical Mediation." *Critical Inquiry* 42:124–48.

Grüter, Christoph, M. Sol Balbuena, and Walter M. Farina. 2008. "Informational Conflicts Created by the Waggle Dance." *Proceedings of the Royal Society* B 275:1321–27.

Grüter, Christoph, and Walter M. Farina. 2008. "The Honeybee Waggle Dance: Can We Follow the Steps?" *Trends in Ecology and Evolution* 24:242–47.

Gu, Qinglong L., Norman H. Lam, Michael M. Halassa, and John D Murray. 2020. "Circuit Mechanisms of Top-Down Attentional Control in a Thalamic Reticular Model." Preprint https://doi.org/10.1101/2020.09.16.300749.

Guenther, Frank H., and Marin N. Gjaja. 1996. "The Perceptual Magnet Effect as an Emergent Property of Neural Map Formation." *Journal of the Acoustical Society of America* 100:1111–21.

Guillory, John. 2010. "Genesis of the Media Concept." *Critical Inquiry* 36:321–62.

Haakenson, Chelsea M., Farrah N. Madison, and Gregory F. Ball. 2019. "Effects of Song Experience and Song Quality on Immediate Early Gene Expression in Female Canaries (*Serinus canaria*)." *Developmental Neurobiology* 79:521–35.

Haig, David. 2020. *From Darwin to Derrida: Selfish Genes, Social Selves, and the Meanings of Life*. Cambridge, MA: MIT Press.

Haladjian, Harry Haroutioun, and Carlos Montemayor. 2014. "On the Evolution of Conscious Attention." *Psychonomic Bulletin and Review* 22:595–613.

Halassa, Michael M., and Sabine Kastner. 2017. "Thalamic Functions in Distributed Cognitive Control." *Nature Neuroscience* 20:1669–79.

Hall, Michelle L. 2006. "Convergent Vocal Strategies of Males and Females Are Consistent with a Cooperative Function of Duetting in Australian Magpie-Larks." *Behavior* 143:425–49.

———. 2004. "A Review of Hypotheses for the Functions of Avian Duetting." *Behavioral Ecology and Sociobiology* 55:415–30.

Hall, M. L., and R. D. Magrath. 2000. "Duetting and Mate-Guarding in Australian Magpie-Larks (*Grallina cyanoleuca*)." *Behavioral Ecology and Sociobiology* 47:180–87.

Hall, Michelle L., and Robert D. Magrath. 2007. "Temporal Coordination Signals Coalition Quality." *Current Biology* 17:R406–7.

Hammer, Martin. 1993. "An Identified Neuron Mediates the Unconditioned Stimulus in Associative Olfactory Learning in Honeybees." *Nature* 366:59–63.

Hanlon, Roger T., and John B. Messenger. 2018. *Cephalopod Behaviour*. 2nd ed. Cambridge: Cambridge University Press.

Hara, Erina, Miriam V. Rivas, James M. Ward, Kazuo Okanoya, and Erich D. Jarvis. 2012. "Convergent Differential Regulation of Parvalbumin in the Brains of Vocal Learners." *PLoS ONE* 7:e29457.

Haraway, Donna J. 2016. *Staying with the Trouble: Making Kin in the Chthulucene*. Durham, NC: Duke University Press.

Harman, Graham. 2011. *The Quadruple Object*. Winchester, UK: Zero.

———. 2002. *Tool-Being: Heidegger and the Metaphysics of Objects*. Chicago: Open Court.

Heidegger, Martin. 1962. *Being and Time*. Trans. John Macquarrie and Edward Robinson. New York: Harper & Row.

Hoffmann, Susanne, Lisa Trost, Cornelia Voigt, Stefan Leitner, Alena Lemazina, Hannes Sagunsky, Markus Abels et al. 2019. " Duets Recorded in the Wild Reveal That Interindividually Coordinated Motor Control Enables Cooperative Behavior." *Nature Communications*. https://doi.org/10.1038/s41467-019-10593-3.

Hoffmeyer, Jesper, and Frederik Stjernfelt. 2016. "The Great Chain of Semiosis: Investigating the Steps in the Evolution of Semiotic Competence." *Biosemiotics* 9:7–29.

Hölldobler, Bert, and Edward O. Wilson. 2008. *The Superorganism: The Beauty, Elegance, and Strangeness of Insect Societies*. New York: Norton.

Hourcade, Benoît, Thomas S. Muenz, Jean-Christoph Sandoz, Wolfgang Rössler, and Jean-Marc Devaud. 2010. "Long-Term Memory Leads to Synaptic Reorganization in the Mushroom Bodies: A Memory Trace in the Insect Brain?" *Journal of Neuroscience* 30:6461–65.

Howard, Scarlett R., Aurore Avarguès-Weber, Jair E. Garcia, Andrew D. Greentree, and Adrian G. Dyer. 2019. "Numerical Cognition in Honeybees Enables Addition and Subtraction." *Science Advances* 5:eaav0961.

Hrncir, Michael, Camila Maia-Silva, Sofia I. McCabe, and Walter M. Farina. 2011. "The Recruiter's Excitement — Features of Thoracic Vibrations during the Honey Bee's Waggle Dance Related to Food Source Profitability." *Journal of Experimental Biology* 214:4055–64.

Ibbotson, M. R. 2001. "Evidence for Velocity-Tuned Motion-Sensitive Descending Neurons in the Honeybee." *Proceedings of the Royal Society B* 268:2195–2201.

Jarvis, Erich D. 2004. "Brains and Birdsong." In *Nature's Music: The Science of Birdsong*, ed. Peter Marler and Hans Slabbekoorn, 226–71. Amsterdam: Elsevier.

Jozet-Alves, C., M. Bertin, and N. S. Clayton. 2013. "Evidence of Episodic-Like Memory in Cuttlefish." *Current Biology* 23:R1033–35.

Kauffman, Stuart A. 1995. *At Home in the Universe: The Search for Laws of Self-Organization and Complexity*. New York: Oxford University Press.

———. 2019. *A World beyond Physics: The Emergence and Evolution of Life*. Oxford: Oxford University Press.

———. 1993. *The Origins of Order: Self-Organization and Selection in Evolution*. New York: Oxford University Press.

Keller, Evelyn Fox. 2010. "It Is Possible to Reduce Biological Explanations to Explanations in Chemistry and/or Physics." In *Contemporary Debates in Philosophy of Biology*, eds. Francisco J. Ayala and Robert Arp, pp. 19–31. Oxford: Wiley-Blackwell.

King, Andrew P., Meredith J. West, and David J. White. 2002. "The Presumption of Sociality: Social Learning in Diverse Contexts in Brown-Headed Cowbirds (*Molothrus ater*)." *Journal of Comparative Psychology* 116:173–81.

Kirschner, Marc W., and John C. Gerhart. 2010. "Facilitated Variation." In *Evolution: The Extended Synthesis*, ed. Massimo Pigliucci and Gerd B. Müller, 253–80. Cambridge, MA: MIT Press.

———. 2005. *The Plausibility of Life: Resolving Darwin's Dilemma*. New Haven, CT: Yale University Press.

Knapska, Ewelina, and Leszek Kaczmarek. 2004. "A Gene for Neuronal Plasticity in the Mammalian Brain: Zif268/Egr-1/NGFI-A/Krox-24/TIS8/ZENK?" *Progress in Neurobiology* 74:183–211.

Knudsen, Eric I. 2020. "Evolution of Neural Processing for Visual Perception in Vertebrates." *Journal of Comparative Neurology* 528:2888–901.

———. 2007. "Fundamental Components of Attention." *Annual Review of Neuroscience* 30:57–78.

———. 2018. "Neural Circuits that Mediate Selective Attention — A Comparative Perspective." *Trends in Neuroscience* 41:789–805.

Kockelman, Paul. 2013. *Agent, Person, Subject, Self: A Theory of Ontology, Interaction, and Infrastructure*. Oxford: Oxford University Press.

———. 2015. "Four Theories of Things: Aristotle, Marx, Heidegger, and Peirce." *Signs and Society* 3:153–92.

Kohn, Eduardo. 2013. *How Forests Think: Toward an Anthropology beyond the Human*. Berkeley: University of California Press.

Krauzlis, Richard J., Amarender R. Bogadhi, James P. Herman, and Anil Bollimunta. 2018. "Selective Attention without a Neocortex." *Cortex* 102: 161–75.

Kroeber, A. L., and Clyde Kluckhohn. 1952. *Culture: A Critical Review of Concepts and Definitions*. Cambridge, MA: Peabody Museum.

Kull, Kalevi. 2009. "Vegetative, Animal, and Cultural Semiosis: The Semiotic Threshold Zones." *Cognitive Semiotics* 4:8–27.

Labhart, Thomas, and Eric P. Meyer. 2002. "Neural Mechanisms in Insect Navigation: Polarization Compass and Odometer." *Current Opinion in Neurobiology* 12:707–14.

Lachlan, R. F., L. Verhagen, S. Peters, and C. ten Cate. 2010. "Are There Species-Universal Categories in Bird Song Phonology and Syntax? A Comparative Study of Chaffinches (*Fringilla coelebs*), Zebra Finches (*Taenopygia guttata*), and Swamp Sparrows (*Melospiza georgiana*)." *Journal of Comparative Psychology* 124:92–108.

Laland, Kevin N. 2017. *Darwin's Unfinished Symphony: How Culture Made the Human Mind*. Princeton, NJ: Princeton University Press.

Laland, Kevin N., and Bennett G. Galef, eds. 2009. *The Question of Animal Culture*. Cambridge, MA: Harvard University Press.

Laland, Kevin N., and William Hoppitt. 2003. "Do Animals Have Culture?" *Evolutionary Anthropology* 12:150–59.

Laubichler, Manfred D., and Jürgen Renn. 2015. "Extended Evolution: A Conceptual Framework for Integrating Regulatory Networks and Niche Construction." *Journal of Experimental Zoology B: Molecular and Developmental Evolution* 324:565–77.

Leadbeater, Ellouise, and Lars Chittka. 2007. "Social Learning in Insects — From Miniature Brains to Consensus Building." *Current Biology* 17:R703–13.

Leboucher, Gérard, Eric Vallet, Laurent Nagle, Nathalie Béguin, Dalila Bovet, Frédérique Hallé, Tudor Ion Draganoiu et al. 2012. "Studying Female Reproductive Activities in Relation to Male Song: The Domestic Canary as Model." *Advances in the Study of Animal Behavior* 44:183–223.

Le Guin, Ursula K. 2016. *The Unreal and the Real: The Selected Short Stories of Ursula K. Le Guin*. New York: Saga Press.

Leitner, Stefan, and Clive K. Catchpole. 2002. "Female Canaries That Respond and Discriminate More between Male Songs of Different Quality Have a Larger Song Control Nucleus (HVC) in the Brain." *Journal of Neurobiology* 52:294–301.

_____. 2004. "Syllable Repertoire and the Size of the Song Control System in Captive Canaries (*Serinus canaria*)." *Journal of Neurobiology*. 60:21–27.

Leitner, Stefan, Cornelia Voigt, Reinhold Metzdorf, and Clive K. Catchpole. 2005. "Immediate Early Gene (ZENK, Arc) Expression in the Auditory Forebrain of Female Canaries Varies in Response to Male Song Quality." *Journal of Neurobiology* 64:275–84.

Lewontin, Richard. 2000. *The Triple Helix: Gene, Organism, and Environment*. Cambridge, MA: Harvard University Press.

Linksvayer, T. A., J. H. Fewell, J. Gadau, and M. D. Laubichler. 2012. "Developmental Evolution in Social Insects: Regulatory Networks from Genes to Societies." *Journal of Experimental Zoology: Molecular and Developmental Evolution* 318:159–69.

Liu, Wan-Chun, Kazuhiro Wada, Erich D. Jarvis, and Fernando Nottebohm. 2013. "Rudimentary Substrates for Vocal Learning in a Suboscine." *Nature Communications* 4:2082.

Luisi, Pier Luigi. 2016. *The Emergence of Life: From Chemical Origins to Synthetic Biology*. 2nd ed. Cambridge: Cambridge University Press.

Mackevicius, Emily L., Michael T. L. Happ, and Michale S. Fee. 2020. "An Avian Cortical Circuit for Chunking Tutor Song Syllables into Simple Vocal-Motor Units." *Nature Communications* 11, no. 1:5029.

Malthus, Thomas. 2008 [1798]. *An Essay on the Principle of Population*, ed. Geoffrey Gilbert. Oxford: Oxford University Press.

Marler, Peter. 2000. "Origins of Music and Speech: Insights from Animals." In *The Origins of Music*, ed. Nils L. Wallin, Björn Merker, and Steven Brown, 31–48. Cambridge, MA: MIT Press.

Marler, Peter, and Susan Peters. 1977. "Selective Vocal learning in a Sparrow." *Science* 198:519–21.

Marler, Peter, and Hans Slabbekoorn. 2004. *Nature's Music: The Science of Birdsong*. Amsterdam: Elsevier.

Martins, Pedro Tiago, and Cedric Boeckx. 2020. "Vocal Learning: Beyond the Continuum." *PLoS Biology*. https://doi.org/10.1371/journal.pbio.3000672.

Mäthger, L. M., E. J. Denton, N. J. Marshall, and R. T. Hanlon. 2009. "Mechanisms and Behavioural Function of Structural Coloration in Cephalopods." *Journal of the Royal Society Interface* 6:S149–63.

Maturana, H. R., and F. J. Varela. 1980. *Autopoiesis and Cognition: The Realization of the Living*. Dordrecht: Reidel.

Maynard Smith, John. 2000. "The Concept of Information in Biology." *Philosophy of Science* 67:177–94.

Maynard Smith, John, and Eörs Szathmáry. 1995. *The Major Transitions in Evolution*. Oxford: Oxford University Press.

McGhee, George. 2007. *The Geometry of Evolution: Adaptive Landscapes and Theoretical Morphoscapes*. Cambridge: Cambridge University Press.

McMillan, Neil, Marc T. Avey, Laurie L. Bloomfield, Lauren M. Guillette, Alison H. Hahn, Marisa Hoeschele, and Christopher B. Sturdy. 2017. "Avian Vocal Perception: Bioacoustics and Perceptual Mechanisms." In *Avian Cognition*, ed. Carel ten Cate and Susan D. Healy, 270–95. Cambridge: Cambridge University Press.

Mello, C. V., D. S. Vicario, and D. F. Clayton. 1992. "Song Presentation Induces Gene Expression in the Songbird Forebrain." *Proceedings of the National Academy of Science* 89:6818–22.

Menzel, Randolf. 1999. "Memory Dynamics in the Honeybee." *Journal of Comparative Physiology* 185:323–40.

———. 2019. "The Waggle Dance as an Intended Flight: A Cognitive Perspective." *Insects* 10:424.

Menzel, Randolf, and Martin Giurfa. 2001. "Cognitive Architecture of a Mini-Brain: The Honeybee." *Trends in Cognitive Sciences* 5:62–71.

Messenger, J. B. 1977. "Prey-Capture and Learning in the Cuttlefish *Sepia*." *Symposia of the Zoological Society of London* 38:347–76.

Michelsen, Axel. 2003. "Signals and Flexibility in the Dance Communication of Honeybees." *Journal of Comparative Physiology A* 189:165–74.

Millikan, Ruth Garrett. 2017. *Beyond Concepts: Unicepts, Language, and Natural Information*. Oxford: Oxford University Press.

———. 1989. "Biosemantics." *The Journal of Philosophy* 86:281–97.

———. 1996. "Pushmi-pullyu Representations." In *Mind and Morals*, ed. L. May and M. Friedman, 145–61. Cambridge, MA: MIT Press.

———. 2004. *Varieties of Meaning*. Cambridge, MA: MIT Press.

Mooney, Richard, Jonathan Prather, and Todd Roberts. 2008. "Neurophysiology of Birdsong Learning." In *Learning and Memory: A Comprehensive Reference*, ed. H. Eichenbaum, 4 vols. 3:441–74. Oxford: Elsevier.

Moorman, Sanne, Claudio V. Mello, and Johan J. Bolhuis. 2011. "From Songs to Synapses: Molecular Mechanisms of Birdsong Memory." *Bioessays* 33:377–85.

Mori, Chihiro, Wan-chun Liu, and Kazuhiro Wada. 2018. " Recurrent Development of Song Idiosyncrasy without Auditory Inputs in the Canary, an Open-Ended Vocal Learner." *Nature Scientific Reports* 8:8732.

Nagel, Thomas. 1974. "What Is It Like to Be a Bat?" *Philosophical Review* 83:436–50.

Nani, Andrea, Jordi Manuello, Lorenzo Mancuso, Donato Liloia, Tommaso Costa, and Franco Cauda. 2019. "The Neural Correlates of Consciousness and Attention." *Frontiers in Neuroscience* 13:1169.

Neander, Karen. 1991. "Functions as Selected Effects." *Philosophy of Science* 58:168–84.

———. 2020. "Teleological Theories of Mental Content." *The Stanford Encyclopedia of Philosophy*, ed. Edward N. Zalta. https://plato.stanford.edu/archives/win2020/entries/content-teleological/.

Newman, Stuart A. 2010. "Dynamical Patterning Modules." In *Evolution: The Extended Synthesis*, ed. Massimo Pigliucci and Gerd B. Müller, 281–306. Cambridge, MA: MIT Press.

Newman, Stuart A., and Ramray Bhat. 2009. "Dynamical Patterning Modules: A 'Pattern Language' for Development and Evolution of Multicellular Form." *International Journal of Developmental Biology* 53:693–705.

Nicholson, Daniel J., and John Dupré, eds. 2018. *Everything Flows: Towards a Processual Philosophy of Biology*. Oxford: Oxford University Press.

Noad, Michael J., Douglas H. Cato, M. M. Bryden, Micheline-N. Jenner, and K. Curt S. Jenner. 2000. "Cultural Revolutions in Whale Songs." *Nature* 408:537–38.

Nordström, Karin, and David C. O'Carroll. 2009. "Feature Detection and the Hypercomplex Property in Insects." *Trends in Neurosciences* 32:383–91.

Nottebohm, F., T. M. Stokes, and C. M. Leonard. 1976. "Central Control of Song in the Canary, *Serinus canarius*." *Journal of Comparative Neurology* 165:457–86.

Nowicki, Stephen, and William A. Searcy. 2014. "The Evolution of Vocal Learning." *Current Opinion in Neurobiology* 28:48–53.

Nuxoll, Andrew. 2012. "Episodic Learning." *Online Encyclopedia of the Sciences of Learning*. https://doi.org/10.1007/978-1-4419-1428-6_1362.

Odling-Smee, F. John, Kevin N. Laland, and Marcus W. Feldman. 2003. *Niche Construction: The Neglected Process in Evolution*. Princeton, NJ: Princeton University Press.

Odom, Karan J., Michelle L. Hall, Katharina Riebel, Kevin E. Omland, and Naomi E. Langmore. 2014. " Female Song Is Widespread and Ancestral in Songbirds." *Nature Communications* 5, article 3379.

Okanoya, Kazuo. 2002. "Sexual Display as a Syntactical Vehicle: The Evolution of Syntax in Birdsong and Human Language through Sexual Selection." In *The Transition to Language*, ed. Alison Wray, 46–63. Oxford: Oxford University Press.

Okasha, Samir. 2006. *Evolution and the Levels of Selection*. Oxford: Clarendon Press.

Oyama, Susan. 2000. *The Ontogeny of Information: Developmental Systems and Evolution*. Durham, NC: Duke University Press.

Pahl, Mario, Aung Si, and Shaowu Zhang. 2013. "Numerical Cognition in Bees and Other Insects." *Frontiers in Psychology* 4, article 162.

Pankiw, Tanya. 2004. "Cued In: Honey Bee Pheromones as Information Flow and Collective Decision-Making." *Apidologie* 35:217–26.

Pasteau, Magali, Laurent Nagle, Marie Monbureau, and Michael Kreutzer. 2009. "Aviary Experience Has No Effect on Predisposition of Female Common Canaries (*Serinus canaria*) for Longer Sexy Phrases." *The Auk* 126:383–88.

Patel, Gaurav H., Danica Yang, Emery C. Jamerson, Lawrence H. Snyder, Maurizio Corbetta, and Vincent P. Ferrera. 2015. "Functional Evolution of New and Expanded Attention Networks in Humans." *Proceedings of the National Academy of Science* 112:9454–59.

Payne, Roger S., and Scott McVay. 1971. "Songs of Humpback Whales." *Science* 173:585–97.

Peirce, Charles Sanders. 1994. *The Collected Papers of Charles Sanders Peirce. Electronic Edition*, ed. Charles Hartshorne, Paul Weiss, and A. W. Burks. Charlottesville, VA: InteLex.

———. 1955. *Philosophical Writings of Peirce*, ed. Justus Buchler. New York: Dover.

Peng, Fei, and Lars Chittka. 2017. "A Simple Computational Model of the Bee Mushroom Body Can Explain Seemingly Complex Forms of Olfactory Learning and Memory." *Current Biology* 27:224–30.

Perlman, Mark. 2010. "Traits Have Evolved to Function the Way They Do Because of Past Advantage." In *Contemporary Debates in Philosophy of Biology*, ed. Francisco J. Ayala and Robert Arp, pp. 53–71. Oxford: Wiley-Blackwell.

Peters, John Durham. 2015. *The Marvelous Clouds: Toward a Philosophy of Elemental Media*. Chicago: University of Chicago Press.

Pigliucci, Massimo, and Jonathan Kaplan. 2006. *Making Sense of Evolution: The Conceptual Foundations of Evolutionary Biology*. Chicago: University of Chicago Press.

Pollan, Michael. 2013. "The Intelligent Plant: Scientists Debate a New Way of Understanding Flora." *The New Yorker*. December 23–30.

Pollard, Thomas D., and William C. Earnshaw. 2008. *Cell Biology*. 2nd ed. Philadelphia: Saunders Elsevier.

Price, J. Jordan. 2015. "Rethinking Our Assumptions about the Evolution of Bird Song and Other Sexually Dimorphic Signals." *Frontiers in Ecology and Evolution*. https://doi.org/10.3389/fevo.2015.00040.

Price, Robbie l'Anson, and Cristophe Grüter. 2015. "Why, When and Where Did Honey Bee Dance Communication Evolve?" In *Ballroom Biology: Recent Insights into Honey Bee Waggle Dance Communications*, ed. Roger Schürch, Margaret J. Couvillon, and Madeleine Beekman, 29–35. Lausanne: Frontiers Media.

Prum, Richard O. 2017. *The Evolution of Beauty: How Darwin's Forgotten Theory of Mate Choice Shapes the Animal World — and Us*. New York: Doubleday.

Rek, Pawel. 2018. "Multimodal Coordination Enhances the Responses to an Avian Duet." *Behavioral Ecology* 29:411–17.

Rek, Pawel, and Robert D. Magrath. 2017. "Deceptive Vocal Duets and Multimodal Display in a Songbird." *Proceedings of the Royal Society B*, October 4. doi.org/10.1098/rspb.2017.1774.

———. 2016. "Multimodal Duetting in Magpie-Larks: How Do Vocal and Visual Components Contribute to a Cooperative Signal's Function?" *Animal Behaviour* 117:35–42.

Rendell, Luke, and Hal Whitehead. 2001. "Culture in Whales and Dolphins." *Behavioral and Brain Sciences* 24:309–82.

Ribeiro, Sidarta, Guillermo A. Cecchi, Marcelo O. Magnasco, and Claudio V. Mello. 1998. "Toward a Song Code: Evidence for a Syllabic Representation in the Canary Brain." *Neuron* 21:359–71.

Richerson, Peter J., and Robert Boyd. 2005. *Not by Genes Alone: How Culture Transformed Human Evolution*. Chicago: University of Chicago Press.

Riebel, Katharina, Karan J. Odom, Naomi E. Langmore, and Michelle L. Hall. 2019. "New Insights from Female Bird Song: Towards an Integrated Approach to Studying Male and Female Communication Roles." *Biology Letters* 15 (April 3). http://dx.doi.org/10.1098/rsbl.2019.0059.

Robinson, Gene E., Russell D. Fernald, and David F. Clayton. 2008. "Genes and Social Behavior." *Science* 322:896–900.

Robinson, Gene E., Christina M. Grozinger, and Charles W. Whitfield. 2005. "Sociogenomics: Social Life in Molecular Terms." *Nature Reviews Genetics* 6:257–71.

Rybczynski, Natalia. 2007. "Castorid Phylogenetics: Implications for the Evolution of Swimming and Tree-Exploitation in Beavers." *Journal of Mammalian Evolution* 14:1–35.

Salwiczek, Lucie H., Arii Watanabe, and Nicola S. Clayton. 2010. "Ten Years of Research into Avian Models of Episodic-like Memory and Its Implications for Developmental and Comparative Cognition." *Behavioural Brain Research* 215:221–34.

Schrödinger, Erwin. 2012. *What Is Life? With Mind and Matter and Autobiographical Sketches.* Cambridge: Cambridge University Press.

Seeley, Thomas D. 1985. *Honeybee Ecology: A Study of Adaptation in Social Life.* Princeton, NJ: Princeton University Press.

———. 1995. *The Wisdom of the Hive: The Social Physiology of Honey Bee Colonies.* Cambridge, MA: Harvard University Press.

Seeley, Thomas D. and Susannah C. Buhrman. 1999. "Group Decision Making in Swarms of Honey Bees." *Behavioral Ecology and Sociobiology* 45:19–31.

Seeley, T. D., A. S. Mikheyev, and G. J. Pagano. 2000. "Dancing Bees Tune Both Duration and Rate of Waggle-Run Production in Relation to Nectar-Source Profitability." *Journal of Comparative Physiology* 186:813–19.

Shah, Aridni, Rikesh Jain, and Axel Brockmann. 2020. "Egr-1: A Candidate Transcription Factor Involved in Molecular Processes Underlying Time-Memory." In Giurfa, Riffell, and Chittka, *The Mechanisms of Insect Cognition*, 72–83.

Shannon, Claude E., and Warren Weaver. 1949. *The Mathematical Theory of Communication.* Urbana: University of Illinois Press.

Sheriff, John K. 1994. *Charles Peirce's Guess at the Riddle: Grounds for Human Significance.* Bloomington: Indiana University Press.

Short, T. L. 2007. *Peirce's Theory of Signs.* Cambridge: Cambridge University Press.

Shumaker, Robert W., Kristina R. Walkup, and Benjamin B. Beck. 2011. *Animal Tool Behavior: The Use and Manufacture of Tools by Animals.* Baltimore, MD: Johns Hopkins University Press.

Si, Aung, Mandyam V. Srinivasan, and Shaowu Zhang. 2003. "Honeybee Navigation: Properties of the Visually Driven 'Odometer'." *Journal of Experimental Biology* 206:1265–73.

Simondon, Gilbert. 2017. *On the Mode of Existence of Technical Objects.* Trans. Cecile Malaspina and John Rogove. Minneapolis: University of Minnesota Press.

Slagsvold, Tore, and Karen L. Wiebe. 2011. "Social Learning in Birds and Its Role in Shaping a Foraging Niche." *Philosophical Transactions of the Royal Society B* 366:969–77.

Slessor, Keith N., Mark L. Winston, and Yves Le Conte. 2005. "Pheromone Communication in the Honeybee (*Apis mellifera* L.)" *Journal of Chemical Ecology* 31:2731–45.

Smith, Darren, Jan Wessnitzer, and Barbara Webb. 2008. "A Model of Associative Learning in the Mushroom Body." *Biological Cybernetics* 99:89–103.

Smolker, R. A., Andrew Richards, Richard Connor, Janet Mann, and Per Berggren. 1997. "Sponge Carrying by Dolphins (Delphinidae, *Tursiops* sp.): A Foraging Specialization Involving Tool Use?" *Ethology* 103:454–65.

Srinivasan, M. V., S. Zhang, M. Altwein, and J. Tautz. 2000. "Honeybee Navigation: Nature and Calibration of the 'Odometer'." *Science* 287:851–53.

Staaf, Danna. 2017. *Squid Empire: The Rise and Fall of the Cephalopods*. Lebanon, NH: ForeEdge.

Sterelny, Kim. 2003. *Thought in a Hostile World: The Evolution of Human Cognition*. Malden, MA: Blackwell.

Sterelny, Kim, and Paul E. Griffiths. 1999. *Sex and Death: An Introduction to Philosophy of Biology*. Chicago: University of Chicago Press.

Stevens, Martin. 2013. *Sensory Ecology, Behaviour, and Evolution*. Oxford: Oxford University Press.

Suthers, Roderick A., Eric Vallet, and Michael Kreutzer. 2012. "Bilateral Coordination and the Motor Basis of Female Preference for Sexual Signals in Canary Song." *Journal of Experimental Biology* 215:2950–59.

Suzuki, Toshitaka N., David Wheatcroft, and Michael Greisser. 2016. "Experimental Evidence for Compositional Syntax in Bird Calls." *Nature Communications* 7:10986.

Szathmáry, Eörs. 2015. "Toward Major Evolutionary Transitions Theory 2.0." *Proceedings of the National Academy of Science* 112:10104–11.

Templer, Victoria L., and Robert R. Hampton. 2013. "Episodic Memory in Nonhuman Animals." *Current Biology* 23:R801–806.

Templeton, Christopher N., Çağlar Akçay, S. Elizabeth Campbell, and Michael D. Beecher. 2012. "Soft Song Is a Reliable Signal of Aggressive Intent in Song Sparrows." *Behavioral Ecology and Sociobiology* 66:1503–509.

Thom, Corinna, David C. Gilley, Judith Hooper, and Harald E. Esch. 2007. "The Scent of the Waggle Dance." *PLoS Biology* 5:1862–67.

Tinbergen, Niko. 1951. *The Study of Instinct*. Oxford: Clarendon Press.

Tomasello, Michael. 1994. "The Question of Chimpanzee Culture." In Laland and Galef, *The Question of Animal Culture*, 198–221.

Tomlinson, Gary. 2015. *A Million Years of Music: The Emergence of Human Modernity*. New York: Zone Books.

———. 2018. *Culture and the Course of Human Evolution*. Chicago: University of Chicago Press.

———. 2016. "Sign, Affect, and Musicking before the Human." *boundary 2* 43:143–72.

Trainer, J. M. 1989. "Cultural Evolution in Song Dialects of Yellow-Rumped Caciques in Panama." *Ethology* 80:190–204.

Tulving, Endel. 1972. "Episodic and Semantic Memory." In *Organization of Memory*, ed. E. Tulving and W. Donaldson, 381–402. New York: Academic Press.

———. 2002. "Episodic Memory: From Mind to Brain." *Annual Review of Psychology* 53:12–25.

Vallet, Eric, I. Beme, and Michael Kreutzer. 1998. "Two-Note Syllables in Canary Songs Elicit High Levels of Sexual Display." *Animal Behaviour* 55:291–97.

Vallet, Eric, and Michael Kreutzer. 1995. "Female Canaries Are Sexually Responsive to Special Song Phrases." *Animal Behaviour* 49:1603–10.

Van Gelder, Tim. 1990. "Compositionality: A Connectionist Variation on a Classical Theme." *Cognitive Science* 14:355–84.

Vasas, Vera, and Lars Chittka. 2019. "Insect-Inspired Sequential Inspection Strategy Enables an Artificial Network of Four Neurons to Estimate Numerosity." *iScience* 11:85–92.

von Frisch, Karl. 1967. *The Dance Language and Orientation of Bees*. Cambridge, MA: Harvard University Press.

Waddington, K. D. 1982. "Honey Bee Foraging Profitability and Round Dance Correlates." *Journal of Comparative Physiology* 148:297–301.

———. 2001. "Subjective Evaluation and Choice Behavior by Nectar- and Pollen-Collecting Bees." In *Cognitive Ecology of Pollination*, ed. L. Chittka and J. D. Thomson, 41–60. Cambridge: Cambridge University Press.

Wagner, Günter P. 2014. *Homology, Genes, and Evolutionary Innovation*. Princeton, NJ: Princeton University Press.

Wagner, Günter P., and L. Altenberg. 1996. "Complex Adaptations and the Evolution of Evolvability." *Evolution* 50:967–76.

Wagner, Günter P., and Jeremy Draghi. 2010. "Evolution of Evolvability." In *Evolution: The Extended Synthesis*, ed. Massimo Pigliucci and Gerd B. Müller, 379–400. Cambridge, MA: MIT Press.

Wagner, Günter P., and Gary Tomlinson. 2022. "Extending the Explanatory Scope of Evolutionary Theory: The Origination of Historical Kinds in Biology and Culture." *Philosophy, Theory, and Practice in Biology* doi:10.3998/ptpbio.2095.

Wang, Hongdi, Azusa Sawai, Noriyuki Toji, Rintaro Sugioka, Yukino Shibata, Yuika Suzuki, Yu Ji et al. 2019. "Transcriptional Regulatory Divergence Underpinning Species-Specific Learned Vocalization in Songbirds." *PLoS Biology* doi.org/10.1371/journal.pbio.3000476.

Waters, Kenneth C. 2004. "What Was Classical Genetics?" *Studies in History and Philosophy of Science* 35:83–109.

Whitehead, Hal, and Luke Rendell. 2014. *The Cultural Lives of Whales and Dolphins*. Chicago: University of Chicago Press.

Whiten, Andrew. 2017. "Culture Extends the Scope of Evolutionary Biology in the Great Apes." *Proceedings of the National Academy of Science* 114:7790–97.

Whiten, A., J. Goodall, W. C. McGrew, T. Nishida, V. Reynolds, Y. Sugiyama, C. E. G. Tutin, R. W. Wrangham, and C. Boesch. 1999. "Cultures in Chimpanzees." *Nature* 399:682–85.

Whiten, Andrew, Antoine Spiteri, Victoria Horner, Kristine E. Bonnie, Susan P. Lambeth, Steven J. Shapiro, and Frans B. M. de Waal. 2007. "Transmission of Multiple Traditions within and between Chimpanzee Groups." *Current Biology* 17:P1038–43.

Williams, Raymond. 1976. *Keywords: A Vocabulary of Culture and Society*. Oxford: Oxford University Press.

Wilson, David Sloan, and Elliott Sober. 1989. "Reviving the Superorganism." *Journal of Theoretical Biology* 136:337–56.

Wilson, Edward O. 1975. *Sociobiology: The New Synthesis*. Cambridge, MA: Harvard University Press.

Wrangham, Richard W., W. C. McGrew, Frans B. M. de Waal, and Paul G. Heltine, eds. 1994. *Chimpanzee Cultures*. Cambridge, MA: Harvard University Press.

Wray, Gregory A., Jeffrey Levinton, and Leo Shapiro. 1996. "Molecular Evidence for Deep Precambrian Divergences among Metazoan Phyla." *Science* 274:568–73.

Wright, Sewall. 1932. "The Roles of Mutation, Inbreeding, Crossbreeding, and Selection in Evolution." In *Proceedings of the Sixth International Congress of Genetics*, 355–66. http://www.esp.org/books/6th-congress/facsimile/contents/6th-cong-p356-wright.pdf.

Index

Zone Books series design by Bruce Mau
Image placement and production by Julie Fry
Typesetting by Meighan Gale
Printed and bound by Maple Press